Old Buildings: Conversion and Restoration

Cost-effective solutions and results in practice

Stuart Smellie

Copyright © 2023 Stuart Smellie

Stuart Smellie asserts his moral right to be the author of this book.

All rights reserved. No part of this publication may be reproduced or used in any form by any means, electrical or mechanical including photocopying, recording or by any information storage and retrieval system, without prior written permission from the author.

Whilst all reasonable care has been made in the preparation of this book, neither the author nor the publisher can accept any responsibility for the consequences arising from the use thereof or for the information contained therein.

ISBN: 978-1-3999-4768-8

First published May 2023

Book production by Cornerstone pa services

Printed and bound in China by Round Color Printing

Contents

Acknowledgements 5
Preface 7
Chapter 1: Introduction..9
    General introduction.. 9
    Structure of the book 14
    What this book is and what it is not... 15
    Risks and Rewards 16
    Assumptions 16
    The projects... 17
    A few explanations 18
Chapter 2: Towards cost-effective solutions..23
    Funding and grants.. 23
    UK Value Added Tax (VAT) 24
    Building design, plans and planning 25
    Old versus new and repair versus replace 33
    Materials: new and reclaimed.. 33
    Renewables and other secondary heat sources 39
    Maintenance and running costs: impact of building solutions............................... 47
    Overview 47
Chapter 3: Case studies – Background and descriptions....................................51
    The Mill 52
    The Garage... 59
    The Farmhouse, the Barn and the Byre: overview 60
    The Farmhouse 63
    The Barn 66
    The Byre... 67
    The Farmhouse Extension.. 68
    The Barn Conversion 69

Chapter 4: Case studies – the builds in practice 73
    The Garage 74
    The Mill 76
    The farm complex: common works to the Farmhouse, Barn and Byre 88
    The Farmhouse 96
    The Barn 114
    The Byre 121
    The later Farmhouse Extension 125
    The later Barn Conversion 127

Chapter 5: Overview 137
    Risks and Rewards 138
    UK Value Added Tax (VAT) 139
    Building design, plans and planning 139
    Build specifications 139
    Finding and working with builders.. 140
    Small versus large 141
    Old versus new and repair versus replace 141
    Materials: new and reclaimed 141
    Heating, renewables and other secondary heat sources 143
    Maintenance and running costs: impact of build solutions 149
    Modern versus old approaches to the buildings 150
    Were the intended results achieved? 153

Chapter 6: Results in practice 153
    How much did they all cost? 156
    Where were the savings made? 158
    Our top and bottom 10 decisions 162

Chapter 7: Footnotes: things that happened along the way 171
Glossary 199
List of photographs 203
List of plans 209
References 211
Further reading 213

Acknowledgements

There are numerous people who have helped over the last 35 years, far too many to acknowledge individually. However, I would in particular like to thank:

Shaun and John Taggart, Mark Ross and the late Glen Smith (aka 'The Boys'), and all the people who worked with them over the years, for everything they have done at the Farm; and notably Brian Wheatley (joiner) and Richard Hillary (fabrication joiner).

I would also like to express my thanks to Alan Ravenhill, joiner and ex-serviceman, now hopefully sunning himself in Spain, who was our main joiner in the Mill project, was a huge help sorting out what was left by his predecessor and overseeing the other later works.

Chris Fish, former Durham City conservation planner and now planning consultant, and Martin Roberts (English Heritage) must be thanked for their help and pragmatic approach to the farm complex of buildings some 20 years ago, and Chris also for his help designing the Barn Conversion 16 years later. They are testimony to the fact that working with planners does work.

Our neighbouring farmer David Jackson and his JCB have been, and still are, a huge help in our life at the farm.

I am also grateful to the Bell Ingram Company (Inverness office), JW Woods (Durham) and also my sister-in-law, Katy Driver, for kindly providing some of the photographs for this book, which are better quality than my own.

I am indebted to John Taylor, architect in Crook (County Durham), for producing our plans for the largest set of projects.

I must express my thanks to This is Me Agency for their kind support in gaining government grant funding; to Dr Rebecca Williams Dinsdale, Lifejoy Coach, for advice; to Lucinda Reed of Cornerstone pa services for her tireless help with editing and proofreading the manuscript and for designing the book; and to Antonia Brindle of Get Brindled for her huge help in setting up the website for the book and making you, the reader, aware of its existence.

Finally I must thank my wife Gillian for all her help with the builds and for tolerating my eccentricities. She has also had enormous input into this book, from advice on content and editing to proofreading before Lucinda's expert services.

Preface

The idea for this book came about in 2021 during the Covid-19 lockdown in the United Kingdom and is based on personal building experiences over a period of some 30 years. Apart from a few minor early projects which I ran myself, my wife, Gillian, and I have carried out seven projects between 1994 and 2016; three on the same farm complex ranging from a 20m² extension to a period house renovation of around 260m² on the same site. We still live in one and we sold another in 2014. These projects raised a large number of building issues and this book describes the principles and options available which we used to variable extents in the different projects. These are examined in general in one chapter and described in detail in a second, illustrating the ideas in the different projects as case studies. Some of the things we did were entirely conventional and simply illustrate some building options, while others were rather different. We have benefitted from our experience since the original builds and therefore have examples of the final results, up to 20 or so years after the initial works, which show how our decisions (however controversial a few were) have lasted over time. The decisions we made and solutions we used, together with the tradespeople working with us, were robust and attractive from our own perspective, and as the tradespeople have gone on to do work for most of our friends and acquaintances over the last 20 years, this suggests that others have felt the same. Our costs were between 30 and 50% less than the quoted averages for restoring or converting old buildings and so clearly these decisions appear to have been cost-effective in addition to producing attractive and robust results. The building process does not need to be prohibitively expensive and restoring or converting an old building can be a realistic prospect for those working on a limited budget.

This book is written from a UK perspective and although several of the projects date back many years, the project principles and ideas remain the same. Similarly, whilst UK taxes, buildings authorities and regulations are described as these apply in 2022 rather than at the time of the projects, these will obviously vary between countries and will almost certainly change in the future in the UK. They are not therefore described in any detail, and readers should always take advice on how they apply in their own country before they start their project.

If an owner uses just one of the ideas in this book to achieve a good final result on their budget, or if the book provides interest and enjoyment to the more general reader, it will have achieved its purpose.

Chapter 1: Introduction

General introduction

Restoring old properties is fun, usually, although on a tight budget it can be extremely challenging. Old buildings can be a source of joy, heartbreak and, if not careful, penury. The restoration or conversion of an old building is an expensive labour of love, but it is possible to achieve this successfully without breaking the bank. From our perspective it would have helped if we had had seven-figure salaries, but we did not, so we were restricted to what funds we had available, hence this book. The projects did, however, come with their eccentricities, some of which were astonishing architectural discoveries and others simply amusing experiences along the way. While the main building stages described in this book are aimed at the would-be building restorer or converter, I hope that the architectural discoveries, descriptions, illustrations and anecdotes will be of interest both to the potential builder and to anyone with a love of old buildings and the exploits that go with them. In some places I have referred to regulations, tax regimes and building authorities in Scotland and England although equivalents exist in all UK countries and the building principles should be relevant anywhere.

I have no building qualifications. I am a retired doctor who has over the years tidied up a house in Oxford, upgraded a few flats in Glasgow and then, with my wife, Gillian, carried out a number of projects of very different sizes and styles which cover most of what a would-be builder might be faced with. We converted a derelict former mill in the north of Scotland, managed the finishing parts of a project in County Durham and carried out a small garage conversion on the same site. We then took on a far more demanding project on a large farm building complex, again near Durham, converted further sections of the buildings shortly afterwards, added a small extension to the Farmhouse some years later and then converted one of the completed projects into a separate residential dwelling in 2016. All of these projects were very different and each represents a different type of build, from the very new to the exceedingly old. We did this without having to resort to bread and water!

The last 30 or so years have been a voyage of discovery through buildings of various ages. My parents did a great deal of house-hunting in my childhood, from which I developed an interest in buildings, especially old ones. The basic principles described in this book are as relevant today as they were 30 years ago, with the exception of some materials which have either changed or been introduced and the advent of the internet and of renewables which have brought with them a desire to reduce our carbon footprint. The latter is not easy in an old building with its inherent draughts and walls which are difficult to insulate but, with a little thought, a great deal can be done to achieve this. There is something very satisfying about the whole building process. It is rather like being self-employed: if a computer collapses it needs to be replaced immediately, without time for bureaucracy, as the show must go on or work and income stops. There was plenty of the former in my life as a doctor, whereas there can be a great sense of achievement when starting the day with a wall that needs to be repaired and ending it to find everything completed without going through a committee.

The process is not, however, for the fainthearted. We have had to make a vast number of decisions, source materials from all sorts of strange places, deal in different ways with architects, surveyors and others, negotiate with planners and building control officers and solve both minor and major building questions in all of the trades involved over the course of the projects. While we managed one project from some 300 miles away, the others were carried out when we were either on site or living a stone's throw away. We have made several mistakes and have learned a huge amount from the people who have worked with us; especially 'The Boys' whom we have known for over 25 years. To be fair, we did start out on the major projects, the Mill building and the projects on the farm complex, with some experience gained from smaller earlier ones, although these experiences pale into insignificance against those in the larger projects. However, these smaller projects were important, not only because of the experience they provided but because they increased our confidence and enthusiasm to take on the larger ones.

Old buildings are lovely places to live in. Unfortunately, rodents, bats, spiders and other assorted beasts think so too. I don't mind bats or spiders whatsoever and it is quite mesmerising to sit outside on a warm summer evening with a glass of wine and watch the pipistrelles fluttering around the building as twilight falls. Rodents, however, are a bit of a different issue.

These buildings may conceal a wealth of features, but can also swallow vast sums of money, and there is no end to the amount that can potentially be spent. Property restoration is also a time-consuming process. In our case, we carried out all of the projects whilst I was working full-time as a doctor. It is amazing what can be achieved with site visits before the working day starts, either to debrief on works carried out the previous day or to plan for the day ahead. This was obviously not possible in the one project we managed from far away.

In my personal opinion, and that of the planning officers involved in our Listed buildings, the optimal solution for an old building is to do as little as possible that affects its fabric: removing the minimum of internal structures, introducing the least number of new wall openings, replacing like with like and not covering up architecturally interesting features.

This book looks at all of the stages of the building process from a range of projects, from design to the finishing groundworks. Six of the projects were on a Listed building or Listed complex of buildings, which in terms of regulatory requirements are more stringent than for other buildings. In practice, and to the admitted embarrassment of the English Heritage officer who visited the site, one of the buildings on the farm complex turned out to date back to approximately 1275 and not the 18th century as it appeared. This building revealed a wealth of features dating from and after the original medieval build and are of historical interest in their own right. In this book, however, they are described in terms of the additional issues they raised and our solutions to preserve the further heritage of the building which emerged without spending a fortune.

One theme which will come up repeatedly is the standard we aimed for in the various parts of the projects. The law of diminishing returns applies here: by simple analogy with wine, a small increase in what you pay for run-of-the-mill plonk will buy a good wine, but not everyone can afford Haut-Brion. Ignoring approaches which could simply be described as cowboy building, we encountered a range of solutions, which can, from our experience, be divided for simplicity into three levels:

1. The gold standard, offered for example by an architect, which could

be described as the most watertight solution possible, but may be extremely expensive.

2. The midway solutions, which cost slightly more than the bare minimum but a fraction of the gold, are well above the regulatory minimum, last the course of time and are pleasing to the eye. They can also make projects easier as the work may be less complex and a significant proportion can be completed by a general builder rather than dedicated specialist tradespeople (the stonemason, joiner, plumber, electrician and painter/decorator etc.). This can avoid delays during the work.

3. The bare minimum, which just meets regulatory requirements but may not stay the course of time and may not be visually particularly pleasing.

The costs of the different options above can differ by an order of magnitude. We found that, in most situations, the second of these solutions was the most cost-effective and produced as good, and sometimes better, results than more expensive approaches.

We also found that general builders who came with a range of skills could carry out many works in these midway solutions. It does not take a specialist, for example, to lay plumbing pipework although it needs a qualified person to complete other tasks such as installing boilers or electrical switches, sockets and mains connections. This book starts from the premise that the average costs of a barn conversion, and also in our case restoration of a period farmhouse, are commonly quoted as being somewhere between £1,500 and £2,000/m² of floor space, although there will be huge variation around this average as each project is unique. As all of our main projects began with buildings which were structurally intact, our costs would always have been at the lower end of this range. Using a combination of our initial decisions and the advice (mostly from our tradespeople), judicious sourcing of materials and midway solutions between the minimum standard and the gold, we reduced our overall costs to well below half the quoted averages. For us, this made the difference between the restoration or conversion being viable or not.

These decisions were supported by our experiences over the 20 years or so since we carried out our first projects together. Where relevant, I have included before, during and after photographs to illustrate the appearances achieved from our decisions and also, more specifically, to show some of our findings in one building after we discovered that it was 500 years older than was thought, and how we approached the things we found. Obviously, not everyone will discover that their property is that much older than it initially appears, but the issues which arose apply equally to any build and our findings not only enriched the whole experience, but also made the process more challenging. Whilst the projects we carried out were only on three sites, the parts of the works, particularly on the large farm complex, were all separate and form a set of very different projects.

I hope that the issues, problems, difficulties, solutions and anecdotes from our experiences will be both useful and in parts entertaining, for anyone taking on a restoration or conversion project on an old building, regardless of its age, and also for anyone with an interest in old buildings.

The largest of our projects involved a complex of interconnected buildings consisting of a farmhouse, barn and adjoining gin gan and a cow byre. This complex had a combined final floor area of just under 500m^2; although the final barn and gin gan (around 160m^2) and cow byre (around 70m^2) were converted slightly later and were separate projects. This may be larger than the projects most people will take on, but in our case, this was by force of circumstances although we could have chosen not to restore all of the buildings. Size, however, is not relevant except that it forced us to examine a wider range of possible building solutions in more detail as the overall costs were potentially so great. All of the principles, problems and solutions described can be up- or downscaled.

There are two main overall approaches available in restoring an old building which will differ, for example, between a dilapidated period house and a barn:

1. To leave all of the old features in the building intact and create a true period property: expose the walls if they are attractive, open up old fireplaces, create vaulted ceilings if attractive or if architecturally interesting roof purlins or other structural beams are present, and use old materials wherever possible for floors, doors, architraves, skirting boards and other finishing joinery. We took this approach with the Farmhouse and the Barn.

2. To convert the building into a modern house, a house within a house, or an old building, depending on the case, still leaving any original features of the building exposed, but creating an overall modern dwelling: this requires a more delicate approach in order not to undermine the heritage of the building and to preserve its main features so that these blend into the modern appearance. This was our approach

with the Mill, the Byre and the more recent Barn Conversion.

Old buildings tend to be large (with a total initial floor area of around 160m², the barn and gin gan section of the complex were much the same size as two modest-sized modern houses), so space is rarely an issue. With this part of the farm complex project of buildings, we found that the best approach to preserve the heritage of the building was to leave the original spaces as open as possible, even if this reduced the available living area which we could have squeezed out of the building. Rather than partitioning rooms in the Barn, for example, we left it open and used it as three different areas, a small 'snug', a dining area and a larger sitting area.

We had a fabulous experience restoring and converting the buildings in our various projects and made some excellent friends from the tradespeople and specialists who helped us along the way. The savings we made were achieved through a few major decisions and several minor ones. These included how, for example, we used architects and surveyors, selected builders and carried out masonry and roofing works. The larger savings amounted to tens of thousands of pounds each and we also made many smaller savings of several thousand pounds each on materials and other building works which, when added together, amounted to a further very large amount of money. In the Mill project, we made our key decisions mostly on advice from our surveyor, and in the farm complex project of buildings on advice directly from our builders. None of these raised any significant building control issues and we did not, in our opinion, cut any corners.

Structure of the book

This book should flow logically. After this introduction, it consists of six further chapters covering:

1. The various options available for a range of renovation and restoration projects, which we used in the case study examples.

2. Descriptions of the projects in the case studies and putting the building projects into context.

3. Descriptions of what we actually did in each project and the effect these had on costs and results.

4. An overview of the results and how the decisions we made influenced these.

5. An examination of whether we achieved what we set out to do in terms of results and savings and our top and bottom ten decisions.

6. A series of anecdotes of things that happened along the way. Some of these carry messages which may be useful

Introduction

while others are simply included because they are interesting or amusing.

What this book is and what it is not

What this book is

This book is a guide to some of the options and solutions available to restore or convert old buildings on a budget. It describes a number of different styles of project and the problems, decisions and range of solutions we encountered or used in order to avoid unnecessary cost to achieve optimal final results and appearances, preserve the heritage of the buildings and create dwellings which have stood the test of time. These are illustrated by the case studies on the various projects we carried out. The book contains a wide range of plans and illustrations before, during and after the works and up to 20 years later. It also describes our various exploits along the way making this a human story of our trials and tribulations.

Main topics examined:
- The range of projects undertaken
- Design and the various consents
- Project specifications and avoiding unnecessary work
- Building options and decisions: the standard versus the less conventional
- Sourcing materials
- Effects of decisions on finished appearances
- Renewable energy and secondary heating
- Mistakes we made
- Cost-effectiveness of the builds
- Top and bottom ten decisions
- Tales and anecdotes during and around the builds.

The main part of the book covers some generalities about the various options to approach a restoration project, a description of our buildings themselves and the case studies describing each of the projects from start to finish illustrating the approaches described in the earlier sections.

In addition to describing our own building experiences, options and solutions, the book examines situations where we went against received wisdom (such as paying daily rates for work rather than obtaining formal quotations) and our reasons for doing this. We have learned some tricks and tips over the years as the projects have escalated from minor house refurbishment and repair to the major building projects.

What this book is not

This book is not a technical guide to building. There are a number of these available, some of which I have listed in the Further reading given at the end of the book. Building techniques are not described except where these are specifically relevant to our decisions on cost, appearance and the ability to last over time (for example, how we pointed the outside walls of the building, a potentially hugely expensive process). The book simply describes our own experiences, which will not necessarily cover all of the issues that may come up during a project.

We are amateur enthusiasts and not developers, and have never restored a property with a view to selling it for profit. Had we had wished to do so we would not have bought some of the properties we did. This is not, therefore, a guide to restoring or converting buildings for profit.

Risks and Rewards

Conversion or restoration of any old building will always bring with it some degree of risk, and decisions about the options at various stages of a project will depend on how risk-averse an owner is. The amount of risk taken will largely depend on who advises on the approaches used during the build process – an architect, surveyor, builder, the owner, or any combination of these.

Budget notwithstanding, some degree of risk taking and judicious shortcuts may be inevitable with an old building. However if the balance of risks and rewards is weighed carefully these can create an equivalent, or often arguably a better, result than more conventional approaches at a far lower cost.

As described later in the book, the decisions we took were mostly mid-range in terms of risk but produced satisfactory results both in terms of appearance, preserving the heritage of the buildings and producing robust and long-lasting results, at the same time reducing costs enormously compared to the highest risk 'gold-plated' solutions,

Assumptions

It is assumed in this book that the would-be builder has some knowledge of building techniques and terms. These can be supplemented by referring to one or more of the guides listed at the end of this book (see Further reading). There is also a short glossary at the end of the book, although this is not exhaustive and only includes terms which a relatively informed owner might not know.

In addition:
- It is assumed that the building has been purchased and that funding is available or has been arranged. The only aspect of funding examined here is a brief description of grant assistance, if only to dismiss this in our own case.
- It is assumed that for any

Introduction

conversion or restoration, the external walls of the building are left in their original state (i.e. not using cladding to enable additional insulation) and that the internal walls are either left exposed, rendered or dry-lined.

- Because VAT rules will vary between buildings, it is assumed that tax will be payable at a rate from 0–20% depending on the project and have given examples of the different rates which would have applied to our different projects in 2022 and the impact these had on some of our decisions. For a period building, unless the property has recently been occupied, I have assumed from current VAT rules that the tax is either refundable (for a DIY builder) or chargeable at a lower 5% rate for VAT-registered contractors on the majority of the build itself, excluding VAT on expert advice when the full rate applies.
- For simplicity it is also assumed, arbitrarily, that around 70% of the actual total potential build costs are for labour and 30% for materials (approximately the case for our projects). Obviously, these figures will vary depending on the build. The labour savings from using a builder who is not registered for VAT are therefore 0–4%, unless the building works attract the standard rate when this saving rises to around 15% (70% of the 20% rate).

- Including VAT, it is assumed that the costs of an architect to manage the project from design through to completion would be approximately 20%, including non-reclaimable VAT. The costs of a planning consultant or architectural draughtsman only to produce and submit the plans for approval, leaving the detailed build specifications and project management stages to the owner, would be around 5%.
- From our experience of quotations from larger contractors and individual tradespeople, a conservative assumption has been made, again somewhat arbitrary and from our own experience, that the additional costs of larger companies compared to single tradespeople (premises, offices staff, directors' salaries and profits/dividends, etc.) are in the region of 25%, although in practice the actual figure may vary considerably.

The projects

The general principles and options are described in Chapter 2 for the projects listed below, which are described in detail in Chapter 3. These case studies cover a wide range of building sizes and types, from around 20–200m², and each of the projects had its own problems, options and solutions. The projects are described from the planning stages through to finished groundworks and the descriptions concentrate on the solutions which had significant effects on the result.

List of illustrative projects:
1. A modern garage conversion project (approximately 50m²).
2. Conversion of a derelict 19th century water mill (approximately 210m²).
3. Restoration of an 18th century farmhouse (approximately 260m²).
4. Conversion of a barn attached to the farmhouse (approximately 160m²).
5. Conversion of a cow byre attached to the barn (approximately 70m²).
6. A later extension to the renovated original farmhouse building (approximately 20m²).
7. Conversion of the barn listed above into a separate dwelling (approximately 190m²).

A few explanations
Property names

To avoid confusion, and this should become more obvious further on, the projects themselves (Gin Gan, Mill, Garage, Farmhouse, Barn, Byre, Farmhouse Extension and Barn Conversion) are capitalised. This is because the book refers, for example, to two gin gans; the early Gin Gan project, which I have only described very briefly, and the gin gan part of the Barn (a gin gan is a structure found particularly, but not exclusively, in the north east of England and consists of a number of stone pillars forming a hexagonal or similar shape, with a roof and internal wooden timbers to support the millstones for a horse-drawn mill). The book also refers to two 'barns'. The project referred to as the Barn includes both the original derelict barn and gin gan part of the buildings making up the farm complex, which together formed a single project carried out shortly after the Farmhouse restoration. The Barn Conversion refers to the later conversion of the initial Barn project, which we turned into a separate dwelling in 2016. Likewise the book refers to several extensions. The term Farmhouse Extension refers specifically to the project which involved building an additional room to the Farmhouse.

Some of the descriptions and plans are common to the different parts of the farm complex projects as these were designed together and were contained in the same plans, even though they were different types of work and were carried out as separate projects.

Other terms

I have used the term 'contractor' to refer to any company, partnership or other provider of services for a build which is registered for UK Value

Added Tax (VAT). I have used the term 'individual tradespeople' to describe one (occasionally more than one) tradesperson or seller of materials working below the VAT threshold, for whom therefore VAT is not payable.

Costs

I have given the costs of materials or works as modern-day equivalents, using the approximate retail price index changes as follows:

1. a factor of approximately 2 since 1994 (Mill, Gin Gan and Garage),
2. a factor of 1.7 approximately since 2000–2004 (Farmhouse, Barn and Extension), and
3. a factor of approximately 1.2 since 2016 (Barn Conversion).

Where costs are compared to reported averages (using for example Checkatrade[1] and other websites), these averages conceal an enormous range, with for example, estimations quoted for external pointing of a wall ranging from £40/m² – £160/m². We did not measure the time taken for individual parts of the works as this would not realistically have been possible. When we worked directly with tradespeople on an hourly or day rate we did not have fixed quotations for the works and therefore have no means of comparing the day rate costs against quotations and we can only compare the final cost to the estimated averages described above. All of the figures quoted are VAT-inclusive where VAT applies although, as discussed in Chapter 2, the actual VAT situation of a project may vary greatly depending on the project from 0–20%.

I have estimated all costs as closely as possible. For some of these it is difficult to give an exact breakdown as, for example, we only have a record of the hours charged over weekly periods, when work would usually be carried out at the same time on several separate parts of the project. I can, however, fairly accurately estimate the savings made in terms of fees for experts, those made from using reclaimed materials rather than new, and those from avoided works or using different building techniques.

Where I have included modern-day costs for specific items or processes such as materials or installation, these are estimates and/or those found from research; please do not factor these into any budget without obtaining quotations or seeking approximate costs from a builder or installer.

Authorities and regulations

The authorities and regulations referred to in this book relate to Scotland and England for the different projects. Similar authorities and regulations exist in other UK countries. Equivalent authorities and regulations will vary between other countries, and readers outside of the UK will obviously need to refer to what applies in their own country. The broad principles, however, are likely to be similar.

Brand names

Apart from the few products which I have listed by name as we found them particularly useful, when I give product names, manufacturers or websites, this is not an endorsement or criticism of any of these. They are purely examples of what we actually used, or did not use, during our projects, or internet sources used for further information such as comparison websites.

Introduction

The guidance in this book is based on our personal experiences or research and, inevitably, some is subjective. I have therefore preceded much of the guidance by the word 'probably'. This is not because I am sitting on the fence but because, in the great majority of cases, when it comes to restoring or converting an old building there are very few right or wrong answers and I would be cautious about any information sources which claim that there are. Also, take account of what could be described as conflicts of interest: most information from companies producing materials, for example, will recommend a technique which uses that type of material, and likewise, few architects will recommend building solutions which do not make use of an architect.

There are often several approaches or building techniques available to achieve the same result, even if some of these go against, the solutions which may be widely recommended. The solution chosen will depend on the individual, based on factors such as the desired final appearance, attitude towards risk and occasionally going against received wisdom, confidence in taking on a project and desired level of involvement in the project (from design through to managing the works themselves), the amount of money available and knowledge of the building processes themselves.

In terms of savings during the builds themselves, apart from the few instances when we had costs to compare to (such as the results of the tender procedure for the Mill and our actual costs described further on) we do not have actual potential costs to compare against our own in practice. The estimated savings reported must therefore be interpreted with caution.

Chapter 2: Towards cost-effective solutions

This chapter describes some of the options which can have a major impact on the final cost and build result, several of which will not necessarily be recognisable from detailed building guides. The case studies in Chapter 4 will illustrate how we used these options to best effect in order to optimise both the robustness of the builds, preserve the heritage of the buildings and achieve the best possible final appearances. Where building decisions and works were straightforward, I have only given a brief description for completeness.

Topics considered:
- Funding and grants
- UK Value Added Tax (VAT)
- Building design, plans and planning
- Build specifications
- Finding and instructing builders
- Old versus new and repair versus replace
- Fabrication joiners
- Materials: new and reclaimed
- Heating, renewables and other secondary heat sources
- Maintenance and running costs: impact of build solutions
- Modern versus old approaches to the buildings

Funding and grants

As explained earlier, I have not considered funding in this book apart from a brief discussion on grant support.

Although grant funding may be available for conversion or restoration of an old building, regardless of what may be available some works may understandably be deemed low priority compared to mandatory grants, such as those to people living in homes with no inside bathrooms. One stipulation we encountered was that the works in question could not be started until the grant was available, bringing with it an unpredictable and potentially expensive delay to the start of works. Some of the stipulations relating to who could or could not carry out particular works also made these potentially more expensive and would have reduced or negated the value of the grant. The whole area of grants is constantly changing and owners should check the grants available at the time and the conditions attached to them before applying for any grant assistance. It is probably better to consider grants as useful windfalls should they be available and not to factor them into the budget.

UK Value Added Tax (VAT)

A detailed description of the VAT rules applicable to building renovation or conversion projects is beyond the scope of this book. As our properties were zero-listed for VAT at the time of the builds, few VAT issues arose as we could reclaim the tax paid for materials and builders. The rules have since changed.

The current VAT rules on building works in the UK are complex. At the time of writing, much of the VAT can be reclaimed by self-builders or charged at 5% (on works agreed with a VAT-registered contractor) for 'qualifying' works on many residential building conversions or the restoration of properties which have not recently been occupied. Detailed lists of what constitute qualifying works for self-builders are available on government websites and on VAT Notice 708 for contractors. The full VAT rate (currently 20%) applies to works which do not qualify; such as restoring a property which has been lived in recently or payments to architects or other building specialists. This applied to two of our projects as detailed in the case studies in Chapter 4 and we therefore benefitted in these cases from using tradespeople who were not VAT registered for these projects.

As standard rate VAT in the UK will add £40,000 to a £200,000 build, this is a huge potential cost consideration. Even at the current 5% reduced rate, £10,000 is still a significant amount of money.

Some of the VAT rules are open to interpretation and appear to vary between works carried out on a self-build basis and those carried out by a contractor and I would strongly advise seeking some specialist advice. This can be obtained either from an architect, if one is instructed to provide services for part or all of the works, or from a building tax specialist, ideally before buying the property in order to budget for the project. Money spent obtaining the correct VAT advice is money extremely well spent, as discovering that a building does not qualify for beneficial tax treatment or failure to comply with all of the requirements to the letter is likely to disqualify an owner from being able to reclaim the tax paid or benefit from the lower rate VAT on the build.

Building design, plans and planning

Older houses raise a number of issues which need to be approached sensitively before and during the design, planning and building processes. Most fundamental is the fabric of the building and the roof. The conventional approach to this is to assess the building fabric using a surveyor. Planning and other permissions may also raise significant issues and may require some detailed research before the buying process. Planning before buying is, in principle, a relatively simple process, much of which can be resolved with a few telephone calls or ideally a pre-planning meeting with the local planning department and, where applicable, the person responsible for Listed building consent. We may just have been lucky but we found our planning and building control officers extremely helpful and parted on good terms, to the point where our local planning officer at the time of the original farm complex projects was our first port of call in the later Barn Conversion.

Building design and plans

There are various options in terms of design and drawing up plans which I have detailed below.

At either ends of the spectrum the building can be designed purely by either the owner themselves or by using the services of a specialist, conventionally an architect. This is not an 'either–or' decision and how the building is designed will depend on the building itself, the potential complexity of the design and on the confidence of an owner. Those designing their own property may well want to instruct any of the experts listed below for advice on both design and planning and will almost certainly need them to produce the necessary plans and elevations for planning permission. There is, therefore, no single ideal which covers every case.

Some of these services extend into the detailed specifications, finding builders and supervision of the works. The range of services, in particular what different surveyors, planning consultants and architectural draughtsmen offer will vary greatly; the descriptions offered below are drawn purely from our own experiences.

- Architects

Many experts will recommend using an architect to manage the whole project as the gold standard approach. It is not, however, the only one. Simplistically, the services of an architect can be broken down into three parts, which extend beyond the design, plans and planning stages:

1. Design, drawing up plans and submitting these for planning permission.
2. Preparing a detailed specification document for the works and putting the different parts of this out to tender.
3. Supervising the works themselves.

For a £200,000 build, each stage will cost somewhere in the region of £6,000–£7,000 (including VAT) using a conventional architect's approach. In terms of risk, the works proposed by an architect probably represent the lowest risk option but not necessarily the most cost-effective and certainly not the cheapest. In a more complex project where there are several possible design options, input from an architect at least for the design stage may, however, be invaluable.

In the case studies below, we used the services of architects in two ways and found one to be more cost-effective than the other.

- Surveyors

Many people will consider surveyors mostly in terms of property valuations and to provide reassurance that the property is not actually falling down or riddled with dry rot. In our own experience, however, we found that the services of surveyors can be used in different ways:

1. For valuation and reassurance that the building fabric is sound.
2. To provide some design solutions and to draw up plans and elevations (a service which we did not use).
3. To advise on building solutions, identify those works which are strictly necessary and to provide some staged supervision of the project. This approach may save a considerable amount of money over the cost of using an architect's services for an entire project if the owner is designing and managing the project themselves.

- Architectural draughtsmen and planning consultants

There is a wealth of companies and individuals with a range of different levels of experience and qualifications offering design and plan drawing services. For simpler projects it may be cost-effective for an owner to have an architectural draughtsman work from their design but to use an experienced planning consultant or architect at least to produce the plans and elevations required for planning permission. The owner may choose to produce their own design, possibly using some of the many online examples available, or to pass this process to an architect or planning consultant. We used an experienced planning consultant in one of the larger case studies described in Chapter 4.

Planning permission

The demands of planning permission will depend on the type of building and whether or not it is Listed; with the latter more stringent controls will be in place to obtain the necessary consents. This book only includes lower Listing grades (2 in England or C in Scotland) and owners really need to take specialist advice for buildings with higher Listing grades (such as grade 1 in England or A in Scotland). On the assumption that the owner has made a preliminary approach to the local planning department the plans produced either by an architect or by an architectural draughtsman should, in theory, be compatible with planning consent. The same should apply to Listed building consent.

For planning applications involving additions to the fabric of the initial building, it is worth looking carefully for evidence of any additions or changes to the buildings, either from the past (for example from aerial photographs) or existing additions, however ramshackle these may be. These may be particularly valuable for Listed buildings. It is possible to obtain aerial photographs for many buildings through UK websites such as Historic England[2] or the National Collection of Aerial Photography[3].

1. Planning officers

Although planning officers are involved principally in the initial planning permission stage (and Listed building consent where necessary), questions may well arise during the building process and require additional planning approval or Listed building consent. We found planning and conservation planning officers to be both helpful and constructive. This may have been helped by the fact that our primary aim was not to make any major changes to the fabric of the buildings.

2. Building control officers

Plans drawn up by an architect or architectural draughtsman should, by default, comply with building control requirements as the plans will normally include these from a list of generic technical descriptions used for building control. However, there will often be different building options available to meet these requirements. Some issues may arise during the building process, identified either by the builders themselves or during the building control officer's staged inspections of the works. It is important to remember that building control officers work to a fixed set of rules and do not have the same degree of flexibility afforded to planning officers.

Build specifications

As described above, many of the standard building control requirements will come 'off-the-peg' from a list of standard descriptions, although the owner or builder may choose the detailed technical specifications used to achieve these requirements (such as achieving the necessary level of insulation or requirements for installing damp proof membranes).

These can be defined in two directions:

1. The conventional approach is for an architect (or the owner) to draw up the detailed specifications for the different stages of the works (masonry, first-fix joinery, plumbing, electrics and finishing joinery etc.), put these out to tender and then choose a builder from the tender results based on the quotations received (see finding builders below).

2. The second and less conventional approach is to work in reverse: to find tradespeople or builders and then define the detailed specifications with them, working stepwise through the build and referring, if necessary, to building control officers and taking advice from a building specialist (architect, surveyor or planning consultant) as needed. Clearly, this requires complete mutual trust between the owner and the tradespeople or builders.

The first approach is the lowest risk option but potentially the more expensive. It also means that new quotations will be needed for any changes to the works. The second approach carries more risk, as the owner does not have the security of quotations for the works. However, it is quite possible to use a combination of these approaches through initial quotations in order to assess the performance of a builder. Although not generally recommended by experts, the second approach can be a useful alternative for more complex projects where the initial plans may change during the build as it offers more flexibility; the owner needs to have confidence in their builder (again, see finding builders below). It is worth remembering that experts such as architects will tend to be risk-averse to avoid things coming back to bite them and impact on their professional indemnity insurance, which brings with it a potentially high cost; being prepared to take modest risk may greatly reduce potentially avoidable cost.

Finding and instructing builders

Finding a reliable builder is, to state the glaringly obvious, the absolute key to any project, and the mutual trust which follows from this is utterly essential. We have dealt mostly with individual tradespeople working for themselves, whom we have either known directly or found from personal recommendations rather than through the conventional tendering process for the works and working with tradespeople or companies we did not know. If finding individual tradespeople is the chosen approach to instructing builders rather than through conventional tenders, the starting point for this is personal recommendation from friends and family. From the theory of six degrees of separation that everyone is likely to know any other person in the world, then someone, somewhere, will have experience of good, reliable builders and other tradespeople. There is, of course, no reason why recommended tradespeople cannot be invited to tender for the works. However, on the one occasion we did this using specifications produced by an architect, several of the tradespeople with whom we worked subsequently told us that they did not tender for the work because the original specifications produced by the architect were so complex. Along the lines of personal recommendations, and seeing the result of their work in our own projects, over the last 20 or so years our own tradespeople – one or more of 'The Boys' with whom we worked on many of the larger projects – have carried out works at some time for almost all of our friends, acquaintances and friends of acquaintances, both locally and elsewhere in the country.

- Small versus large

Single tradespeople can bring with them two main cost advantages and one practical one.

As the current VAT threshold is approximately three times the national average wage of approximately £15/hour (based on the median wage of £611/week quoted by the UK Office for National Statistics), it is likely that a single tradesperson who works predominantly on a 'fit only' basis rather than supplying their own materials will fall below this threshold in a given year. Some builders argue that working on a 'fit only' basis leaves them open to the owner providing substandard materials, but this should not occur if there is sensible dialogue and trust between the two and the owner is not trying to carry out a project 'on the cheap'. Insisting on working on a 'supply and fit' basis also, almost by definition, removes or greatly restricts the option of using reclaimed materials.

The second and obvious advantage of single tradespeople is that they are inherently less expensive than larger contractors as they do not have the same overheads (office premises, office staff, directors' payments etc.). It is difficult to estimate the true cost difference between small and larger operators. I have assumed a figure of 25% in this book, which is conservative from our own experience when we had quotations from both individual tradespeople and larger companies in order to be able to compare the two. We are currently charged an hourly rate which is some 20% over the UK national average for the various tradespeople we use and unless a company is paying skilled tradespeople well below the national average wage it is most unlikely that it could recoup this cost difference despite possible economies of scale. Morally, it is only reasonable that a skilled tradesperson is remunerated fairly for the work they do.

A third and practical benefit of using individual tradespeople is that it can be easier to work directly with the person carrying out the work rather than through a company and, as discussed above, obtaining repeated quotations for any changes in the building plans can be troublesome, potentially very expensive and introduce delays.

The main disadvantage of using individual tradespeople is that they may not be able to provide the warranties that larger contractors may offer for the works. This approach does therefore carry some additional risk over instructing a larger contractor.

- Moving builders around the country

If a larger contractor is taken on to carry out the project on an old building it is likely this will be a local company as few, other than large national companies, will be prepared to work many miles away. This may not apply to the same extent to individual tradespeople. In one of our earlier projects, we wrongly assumed that it would not be viable to recruit local tradespeople where we lived and ask them to work some 300 miles away. This was incorrect: our builders in the farm complex projects have worked all over the country. The costs of temporary accommodation, at least until a sufficient amount of the property is habitable, may be far outweighed by the reassurance that the building work will be reliable and cost-effective.

- Quotations versus day rates

This is controversial. Anyone who has read the media or watched the numerous television programmes about building restoration or renovation will be familiar with the mantra that an owner should never offer to pay a builder or other tradesperson by the hour, but to obtain a fixed quotation for the works in question. There are certain works, such as joinery fabrication (windows etc.), installing electrics or central heating, groundworks and decorating, where a fixed quotation may be useful as the works are not necessarily complicated and are unlikely to raise unforeseen problems; although the same principles of finding builders based on personal recommendations still apply. Clearly, it may be wiser to accept a higher quotation from a company or tradespeople with particularly good personal recommendations than simply to take the lowest quotation.

Converting an empty barn or other building is a far more straightforward process than restoring an existing house as, generally, limited structural work will be necessary on the internal building fabric and processes such as laying a damp proof membrane are less complex. Consequently, fewer unforeseen problems are likely to arise, assuming the owner has the reassurance that the fabric of the building is sound, and a fixed quotation may be a preferable way of working, although even in this case changes may be made to the initial works and require further quotations.

However, where the work is complex and unforeseen problems or several changes in building plans are likely to arise along the way, any quotation from a reputable tradesperson or company will have to take the potential problems into account. In simple terms, this means that the quotation may need to be high unless the building contract contains a contingency clause to cover unforeseen circumstances, in which case the quotation is effectively open-ended and essentially meaningless.

A conservative non-risk taker may prefer fixed quotations, so the approach taken will vary between different people. If a job such as digging up a floor to lay a damp proof course proves more difficult than expected, a quotation protects the owner against additional cost. Conversely, if it proves easier than expected the owner will, obviously, not benefit from any saving. If there is a strong bond of trust between the builder or other tradesperson and the owner, paying people by the hour or day can, in our experience, be a perfectly viable alternative compared to obtaining fixed quotations. This obviously carries a degree of risk and, like all things in a conversion or restoration project on an old building, the owner needs to consider the potential risks and benefits carefully. There are few absolutes in this world and this is no exception, apart from the absolute need for mutual trust.

Many people will argue that paying someone by the day is not an incentive to work quickly, but only an unscrupulous tradesperson would work slowly to increase their income. On the only occasion this occurred in our own experience we became aware of the slow progress at an early stage. The works in two of our projects were unpredictable (remedial works from a joiner who had failed to deliver in one project and complex restoration of a period farmhouse in the next). As detailed in Chapter 4, we found this way of working perfectly satisfactory and it allowed us to respond quickly to new findings and change our build specifications with no unpleasant surprises. In practice, and again as detailed in Chapter 4, we continued to use this approach throughout the later four projects.

The converse argument to the lack of incentive for a tradesperson to work quickly is the 'honey with extras' principle; that once a builder or tradesperson has agreed a contract based on a quotation, this gives little incentive to be competitive after they have been awarded the contract in the first place. Using companies or tradespeople based on strong personal recommendations should, at least in theory, get around these issues of incentives.

Neither approach is therefore risk-free, but what can be said with confidence is that the 'fixed quotations' mantra is not necessarily always correct.

A final point to consider is that although it is possible to withhold payment if a contractor or tradesperson is working to a quotation and fixed contract but does not deliver to the owner's satisfaction, legally this becomes extremely messy. Breaking the works down into stages with a performance review can help (although almost by definition these stages will be relatively large and potentially difficult to break down suitably for an individual tradesperson carrying out one part of the works: masonry, first-fix joinery or finishing joinery etc. This still doesn't protect, for example, against a builder or tradesperson giving an excessively high quotation for any change to the works during one of the stages, as it is unlikely that an owner will bring in a second party. Quotations do not, therefore, completely protect against failure to deliver and offer little, if any, protection in the event that the works should change during the project. In the unhappy event that work is not carried out to a client's satisfaction, it is potentially easier to 'sack' the tradesperson being paid by the hour or day at almost any stage, although this could leave some works only part completed. We experienced this in one of our projects.

Old versus new and repair versus replace

We prefer to preserve the old but to quote the famous Scottish duo, the Corries, from the latter decades of the last century, 'if the shoe fits, wear it', and modern approaches to an old building can be very attractive when the old is married well with the new. When new materials are used, however, high-quality offerings such as new pitch pine, Douglas Fir or hardwood, for example, are very, very expensive. The 'knee-jerk' solution to some damaged existing structures may be to replace the old with new, but this can be both hugely expensive and can clash with an otherwise old appearance. External reroofing or internal replacement of floors, staircases and doors, are all good examples of this. These repairs may be more difficult to cost in advance and are another reason to work on a day rate basis rather than a quotation. As described in the case studies, five-figure savings can be made by repairing rather than replacing and can give a more sympathetic final appearance.

- Fabrication joiners

The fabrication joiner is an essential, but often unsung, hero of any building project. He/she can create almost anything from wood, including windows, doors, architraves, staircases and, in our case, even a cupola. These people generally do not venture onto the building site other than to view what is required, but work their magic from a workshop, replicating old structures, often at a fraction of the cost of large specialist companies offering custom-made woodwork. Finding a skilled fabricator is an invaluable preliminary before embarking on a build, particularly an old property.

Materials: new and reclaimed

Reclaimed materials go far beyond the obvious 'low-lying fruit' such as flooring, even if it means traipsing across a sodden farm field to look at an old bath! In many ways, being restricted by budget helps as this demands more imagination: throwing money at a project does not necessarily mean a better result. The various savings that can be made are not always that great, but when added together can have a large impact on total costs and the appearance of the final result. It is worth remembering, however, that some materials may prove unviable (either impossible to use or their use would require unrealistic labour costs) and owners should be very cautious about insisting that a tradesperson uses materials which they feel are not viable.

In many cases reclaimed materials can be significantly less expensive and may also better preserve the heritage of a building, as many new equivalents, especially wood such as floor joists, may need to be covered up to avoid clashing unacceptably with the otherwise older appearance – and in the case of floor joists makes it impossible to leave the underside of floorboards exposed.

Buying reclaimed materials today is quite different to 20 years ago because of the advent of the internet. This has brought with it far more sources for obtaining materials and, in part, a consequential expansion in the courier and shipping industry. In some cases, it has also led to a hike in costs, as the value of some materials is increasingly being recognised. In addition, many demolition companies have diversified into retailing reclaimed materials, which do fit far better in older buildings but can now cost much the same as their new equivalents unless the buyer is careful in their choice of suppliers. There are, therefore, fewer bargains to be had today but there are still, for example, a large number of demolition companies with sources to find things on the internet.

Several of the local free advertising newspaper sources (such as the regional equivalents of the weekly London Loot paper) have now been replaced by websites such as eBay, Facebook Marketplace, Gumtree, Preloved and dedicated local advertising and recycling websites. Many of these offer materials which can be found locally (e.g. Facebook Marketplace, Gumtree, and using the 'nearest first' search option on eBay rather than the 'most relevant' default).

As modern-day delivery costs have fallen significantly in real terms compared to 20 years ago, buying materials remotely can be perfectly viable. These sites include, but are by no means restricted to, AnyVan[4] and Shiply[5]. There are also inexpensive means of delivering pallets for large or heavy materials, such as UK Pallet Commercial Deliveries[6], or price comparison sites, such as ParcelCompare[7]. Some of the national pallet companies quote figures in the region of £60 to transport a standard half pallet (carrying something about the size of a single oven Aga-style range weighing up to around 500kg and about 1m in length). Even after factoring in transport costs, the savings from buying remotely can still be considerable.

Materials considered:
- Stone and pointing materials
- Roofing and guttering
- Windows
- Insulation
- Wood
- Flooring
- Plumbing, heating and electrics
- Finishing joinery
- Doors and door furniture
- Kitchens, worktops and bathrooms

Stone and pointing materials

Many owners will not have much knowledge of masonry so if, for example, stone is required for a renovation/restoration project it is probably safer to leave sourcing to a builder/stonemason. In practice, most people will only require modest amounts of stone unless a project involves extensive building or rebuilding of large areas of wall. Demolition yards and the assorted available websites remain a good source for these and also for additional materials such as stone lintels.

Many old buildings will have been pointed externally and rendered internally using a form of lime mortar, which is widely reported to allow the wall to breathe compared to preparations containing cement. I have discussed this in more detail further on (see Page 90). Pointing materials play a limited part in the overall cost of the process, as the majority of pointing costs are labour, although the materials required can still be extremely expensive. From their experience on a World Heritage Site, which has stringent specifications for lime pointing, our own local tradespeople estimate that the cost of commercial lime putty and the layers of lime mortar needed for internal finishing amount to approximately 3–4 times the cost of a wet lime, cement and sand mix. This can represent a difference of several thousand pounds for an entire building (quite apart from the huge increase in labour cost). In any form of conservation area the local planning or Listed buildings authorities may dictate the choice of pointing materials although this may, however, be open to some negotiation and persuasion.

Roofing and guttering

Depending on planning or Listed building constraints, and assuming there is a roof, if it needs to be replaced in its entirety the cost may be prohibitive making the whole project uneconomic. If only patches of a roof need to be replaced, owners should be aware of the myriad of subtle differences in shape and colour of tiles and slates. However, with a little searching it is usually possible to find at least a few closely matching ones to replace those which are missing or damaged, but sufficient numbers to reroof a large property will be a rare find. In addition, at least in the case of pantiles, new materials cannot replicate the appearance of an old roof, and replacing the roof of part of a building complex with new materials may clash with the appearance of the remaining area. In such cases there may be no other realistic option, in terms of appearance, other than to replace the entire roof. The standard figures quoted for this are typically £100/m^2, or around £20,000 for an average roof, assuming no work is required to the supporting timber structures. This will also result in the usual appearance of many barn conversions: a brand new roof on an otherwise old building, which may or may not be desired. A thorough investigation of the roof is therefore invaluable before embarking on replacement. It may well be possible to avoid extensive roof replacement or replacing the whole thing regardless of what an expert advises.

Windows

In most cases, unless replacing the windows throughout a property, any new windows will have to match the originals (in a house) or need to be as sympathetic as possible (in a barn). It is extremely unlikely that any suitable reclaimed windows will be found and these will normally need to be custom made. I am assuming without further explanation that any reader planning a sympathetic restoration or conversion will not intend to use styles of window such as uPVC or equivalent but will try to replicate the original in both appearance and materials from which they are made.

Insulation

The cost of new insulation will usually run to several thousand pounds depending on type, particularly for some of the thin highly-insulating materials such as polyurethane used to avoid covering up attractive roofing features such as old purlins. Polyurethane or foil/foam materials are essential when only minimal space is available for the insulation. These are not cheap but typically provide over five times the insulation properties for the same thickness of conventional polystyrene or fibreglass. Simple decisions, such as using 'seconds' grade or reclaimed insulation boards can easily cut costs in half or more.

Wood

The wrong choice of wood, however well-intentioned and expensive, can have a disastrous effect on a building. Apart from flooring, which I have discussed separately, it is possible to find excellent quality reclaimed materials from local advertising sources, websites or purely by luck from roadside discoveries. In order not to delay the works these really have to be bought in advance, ensuring that the owner is always at least one step ahead of the builders and materials are available when they come to a particular stage in the works.

A simple illustration of the 'right wood' is replacement floor joists: if left exposed, modern joists will look completely out of place alongside existing timbers and will really need to be covered with plasterboard, making it impossible to leave the joists or undersides of the upper floorboards exposed. This can easily be achieved using reclaimed joists, which can exactly replicate the appearances of those already in place. If these are not found in time, however, the whole building process can potentially be delayed.

Flooring will probably account for the majority of the wood costs for most projects and can be eye-wateringly expensive depending on type and supplier. Excluding chipboard, the cheapest option of whitewood (at around £10/m² – £15/m²) really needs to be covered with carpet or other material as it is difficult to achieve a stained finish in a large area which fits comfortably with the appearance of an old building. The potentially extremely expensive option of new hardwood or good-quality softwood (Douglas Fir for example) costs upwards of £50/m² plus sealing costs if that is needed, or around £10,000 for a 200m² building. Lying between these two options it is not difficult to find reclaimed boards, either old or modern. Apart from the cost saving, the imperfections in reclaimed boards can be aesthetically pleasing.

Inevitably, at some stage in their lives, some boards will need to be lifted and relaid because of a leak or some other reason. Older boards are more forgiving and can be relaid these with a less obvious impact on final appearance. This is not necessarily the case with plush new boards. Likewise, old boards can be scrolled into the edge of exposed stone walls with a less obvious result than new boards when the scrolling can be more visually jarring.

Plumbing, heating and electrics

There are limited options in terms of plumbing, heating and electrics other than finishing materials such as electrical sockets and switches and plumbing sanitary ware. It is definitely worth shopping around for these. Some of the offerings available from, for example, suppliers such as Screwfix, Toolstation or TLC Electrical Supplies for electrics and the many online plumbing and bathroom supply sites are high quality and inexpensive compared to those from specialist suppliers, which potentially increase costs by a factor of five or more.

Finishing joinery

It is relatively easy to use reclaimed wood to create doors, door jambs and other joinery in 'older' look projects. Cutting large amounts of reclaimed wood for skirting boards and architraves is not necessarily the cheapest option considering the labour involved, but it can still cost significantly less than their new equivalents and can provide a far better final appearance. The fabrication joiner (see earlier, Page 33) is invaluable for this.

In terms of doors and door furniture a huge range of different solutions and sources are available, again partly dictated by the intended final appearance of the building. Existing doors, even if damaged, can often be used effectively or replicated from reclaimed wood in the 'older' look builds and respect the heritage of the building better than new doors costing much more. It is also often possible to refurbish existing door furniture successfully, even if it appears beyond redemption, which will also help to respect the original appearance of the building. It is easy to forget the cost of door furniture, but this can be an extremely expensive addition to the doors themselves as specialist offerings can easily cost north of £100/door. Depending on the choice made, their appearance can also clash with the original building regardless of the amount of money spent.

Kitchens and bathrooms

To state the obvious, it is not necessary to buy a kitchen from a specialist company unless the owner wants to spend a large amount of money. In order to keep an old appearance a simpler kitchen may detract less from the overall look than a more complex one and may be far more cost-effective. Keeping the units as far as possible to standard single (500mm) and double (1,000mm) sizes creates a simple (and far less expensive) overall kitchen appearance and can fit better alongside the existing period features in the room.

It is easy to think of kitchen worktops as incidentals, but like door furniture they can cost as much as a simple kitchen itself. Cheaper laminates are probably a false economy in terms of appearance and durability, whilst extremely good, more expensive, laminate offerings are now available. Solid wood worktops cost around twice this, or approximately £100/m^2. These are attractive and relatively easy to maintain (marks and small scratches can simply be sanded and re-oiled, although beware of burn marks). Wood worktops also offer an attractive finish in a converted or restored old building. Composite materials such as Corian® or cheaper equivalents cost in the range of £700/m² – £1,000/m² and are generally custom-made from a template. These are typically supplied in single pieces ready to install. Advantages of composite worksurfaces include that they can be manufactured with, for example, an integral sink and drainer, they are more heat- and stain-resistant and more durable. However, these benefits come at a high price. Stone (marble, granite or stone composites) cost around the same and produce an excellent appearance but can be more difficult to install, particularly if they need to be cut, and can crack, an expensive risk.

Likewise, judicious internet searching can be a source of good-quality, attractive sanitary ware for a fraction of the cost of specialist suppliers.

Renewables and other secondary heat sources

Although not necessarily part of the initial building works, I have examined these in terms of their impact on the future running costs of an old property. We are (hopefully) all now attempting to reduce our carbon footprint and minimise our use of fossil fuels but, for most of us, at a viable cost. Options for renewables in a domestic setting vary considerably in terms of the amount of heat or electricity they can generate and also in their economic viability and payback times.

This book does not consider the various UK government incentive schemes, as these are rapidly moving targets with ever-changing goalposts for what qualifies, who needs to install them, the actual incentives available and how long any ongoing payments will last once the renewables system is in place. However, incentives can completely change the financial argument for the installation of a given form of renewables and Government grants and will need to be examined from the schemes in place at the time of the project. Also, be prepared for some bureaucracy.

The current UK government renewables incentives are included in the Clean Heat Grant/Boiler Upgrade Scheme that aims to help existing small domestic buildings transition to low carbon heating systems for their homes. Good background information can be obtained from the Energy Saving Trust[8] and estimates can obviously be obtained from the companies which install them, although the owner should bear in mind that these estimates are likely to be based on best-case scenarios. I have not included thermal energy or electricity storage systems, and what is described below looks particularly at the cost and expected return on investment in the systems alone. As detailed below, for a given investment in these systems, the short payback time of the air-to-air solution we chose combined with other benefits which are specific to these systems were, for us, the preferred solution (ignoring Government incentives).

Renewable energy is derived from natural processes that are continually replaced. There are a number of methods available to produce renewable heat or electricity in the domestic setting.

This book looks at electricity and heat-producing renewables as they relate in particular to an old building being restored from scratch, i.e. one in which the floors are not present, working from the screed above a damp course membrane, and therefore assuming that floors do not need to be ripped up for underfloor piping when this is used.

All of the available systems potentially reduce a building's carbon footprint, but to variable extents. The case studies and descriptions further on consider their relative financial viability compared to the energy or heat they generate.

It is perhaps useful from environmental terms to bear in mind that:

1. Any system which uses electricity to generate heat (or indeed energy to run a car), including the assorted types of heat pumps, is not carbon neutral as approximately half of the electricity in the UK is currently generated from fossil fuels; although this is planned to fall greatly in the relatively near future.

2. Any system which requires the use of concrete, such as wind turbines (for the foundations), generates almost 1 tonne of CO_2/tonne of cement produced, so these systems are not completely zero carbon.

I have not considered hydroelectric electricity generation, as very few houses will have the option of doing this.

Like any other building, a period house or conversion can be heated by a conventional central heating system with a gas or oil boiler using radiators or underfloor heating. The discussion below assumes that for any system relying on circulating hot water and radiators (as in a typical conventional central heating system) the radiators are fitted with thermostatic radiator valves (TRVs), which can be turned to the low or off position in rooms not in regular use, such as spare bedrooms. It also assumes that, where feasible and depending on the system and layout of rooms, a zone valve is used to isolate separate areas of the house in order that the valve can be programmed to heat the different areas at different times. It also assumes that the system used does not rely on the TRVs being turned up and down on a daily basis as this is both tiresome and, in practice, not likely to be done consistently.

- Renewables

Various forms of renewables are available, some of which generate electricity and some of which generate heat to replace conventional gas or oil boilers or to operate alongside these. Some of the heating types of renewables are not well suited to older, larger properties, although all of the standard options will be described for completeness. Whilst solar and wind electricity sources operate with no power input, heat pumps require electricity to drive the pump. Based on the assumption that around 50% of UK energy is currently produced from fossil fuels and that, simplistically, heat pumps generate approximately 3.5 times the amount of energy than the energy required to run the pump, (although this varies slightly between heat pump type) then only around a sixth of the energy they produce is derived from fossil fuels. This will fall further in the relatively near future as fossil fuel-based electricity generation is replaced by nuclear or renewable sources.

Renewables considered:
- Solar energy
- Wind energy
- Heat pumps: ground-source and air-to-water systems
- Heat pumps: air-to-air systems
- Other heat sources

1. Solar energy

Solar panels generate electricity both for personal use and to feed into the National Grid, currently with a minimal feed-in tariff. Their greatest benefit is perhaps for those people who locked into the higher feed-in tariffs offered as an incentive several years ago. These tariffs were far greater than the retail price of electricity at the time and were therefore economically unviable. Excluding any government incentive, the payback time for solar panels can probably be measured more in terms of a decade or so rather than in years and only a limited amount of the electricity generated will actually be used in the home rather than fed into the National Grid. While they may produce a couple of hundred pounds' worth of savings/year (according to our own current UK Energy Performance

Certificate details), they are not that cheap to install (around £5,000–£7,000) and depreciation must also be factored in (but is usually not in any of the common descriptions of the systems). The panels are estimated to last for 25–30 years and the inverter for around 10 years, so the depreciation of over £200/year will significantly reduce their financial savings.

Historically and at the time we considered them, in order to benefit from these systems the owner needed to use the electricity at the time it was generated. They therefore had to change the timing of when they used electrical appliances unless the power generation could be combined with continuous usage such as for heat pumps, and more recently electric car batteries. This paradigm has changed however with the introduction of improved battery systems allowing the electricity generated to be stored more efficiently for personal use. This has made solar (and wind) systems more cost-effective although requiring substantially more capital investment. However, this is an evolving technology and owners would be well advised to do their sums carefully at the time of a build, taking account of the technology available and government incentives at the time, but not forgetting to factor in likely depreciation.

2. Wind energy

The issue of locking into generous feed-in tariffs also applies to wind energy. Wind turbines can, in principle, be installed for any property with available land where the prevailing wind is not turbulent, because of neighbouring buildings for example (hence, amongst other reasons, their limited use in towns), but may in some cases require the appropriate planning permission. Simple wind turbines fitted to the gable end of a house are, perhaps, rather contentious but there are a couple of these systems available now which may be more viable than the earlier options. The reader can refer to the article available on the Renewable Energy Website[9] for more information. What is not in dispute is that these house-mounted turbines produce limited energy compared to larger land-mounted ones.

Land-mounted turbines create a modest amount of electricity, to use or feed into the National Grid in the same way as solar energy. They are expensive to buy and install (typically around £20,000–£30,000). Although a typical system may be rated as 5kW it is worth remembering that quotations will probably be based on the average wind speed and will also not take account of depreciation. High wind speeds may exceed the cut-off speed of the turbines, so the average wind speed may overestimate potential energy generation and, obviously, calm days or light breezes will generate minimal electricity. Initial wind speed estimates are available from internet databases before buying an aerometer or obtaining figures from installation companies. Like solar generation systems, the electricity generated for personal use is limited and the overall payback time can again be measured more in decades (probably several) rather than in years. With an estimated lifespan of 20 years, depreciation significantly reduces or may negate their economic benefit. Also, like solar electricity generation, the owner will need to use the electricity produced by these systems at the time it is generated, so the same issue arises as described for solar panels.

3. Heat pumps: ground-source and air-to-water systems

Air and ground-source central heating systems using the same circulating water principle as a conventional oil or gas boiler system with radiators or underground pipework have recently come more in vogue, although they have been available for many years. They are generally intended to replace an oil or gas boiler as it is extremely difficult to combine the two in a single system. Ground-source systems use a heat exchanger to capture the heat from underground pipes and transfer it to underfloor pipework or to large radiators. They are rated by their coefficient of performance (COP), which is the amount of heat delivered by the system compared to the energy used to run the pump. A ground-source heat pump COP of 4 therefore means that for every kW of electrical energy used the system produces 4kW of heat. Because these systems provide low level heating from slow water flow rates, long amounts of underfloor piping or large radiators are required in order to deliver enough heat and this works less effectively in old buildings which are inherently more difficult to insulate than modern new builds. The ground-source systems are also not cheap to install (typically £15,000–£30,000 for a complete installation) and, excluding any government incentives, the payback times for most systems will again be well over a decade. Assuming a lifespan of around 30 years, depreciation will be around £500/year–£1,000/year.

The air-to-water systems use an external fan to draw in air, expel it at a lower temperature and then use a heat exchanger to transfer the energy into a water-based system. The smaller, less expensive air-to-water systems can be used to heat water alone, although anyone who has lived in an older property will be well aware that domestic water heating accounts for only a small fraction of the fuel costs of the building and that by far the majority is spent on space heating. Water-based systems intended to provide space heating generally deliver the heat through similar pipework or oversized radiators as the ground-based systems and have the same potential limitations, particularly in old buildings. I have not found systems over 16kW; this is only sufficient to heat a small old property if used as the primary central heating system.

Like ground-source heat pumps, the standard air-to-water heat pumps generally operate near-continuously at lower flow rates to warm a space, do not deliver immediate room heating and are not intended to be switched on and off regularly according to how rooms are used, unlike the air-to-air systems (see below).

High temperature water-based heat pumps are available and circumvent the limitations of the more mainstream lower temperature versions. These may make water-based systems more viable but are not yet in widespread use, are significantly more expensive and consequently have a greater depreciation cost.

Few if any of the high temperature heat pumps have an individual output liable to be sufficient for the typically larger volume of an older property and it would seem unlikely that they could replace a conventional system in these buildings at least at present. It would, however, be well worth checking the situation with a heating engineer at the time of a build as things may change from the situation in 2022.

4. Heat pumps: air-to-air systems

Air-to-air heat pump systems have, at least until now, been the Cinderella amongst renewables and have not had much attention in the media despite being low cost and efficient options. They are a cheaper, and arguably far better, renewables option than the present ground-based or air-to-water systems for older properties. They can generate far more energy than small wind or solar systems and are well suited to older properties, particularly for large spaces, heavily-used rooms and spaces out of practical reach of an existing central heating system such as outside studios. As these are essentially highly-efficient air conditioners working in heating rather than cooling mode (although they can also be used for cooling), they are not complicated to install. They operate on the same principle as air to water systems. Their COP, over 3.5 in mild weather, typically drops to under 2.5 in very low temperatures, and makes them slightly less efficient than ground-source systems. Although they can be used to replace a conventional central heating system this requires more complex multi-head split systems or multiple heat pumps, which is a significantly more complicated and disruptive process than installing single stand-alone units. We have used these systems very effectively alongside a conventional oil central heating system.

It is also simple to switch these on and off regularly, as they deliver warm air at the time it is needed for an individual space rather than to a larger part of a house. They can comfortably raise the temperature of an average room quickly to over 25°C and can be particularly effective in heavily used rooms when they can greatly reduce the need for a conventional central heating system. This is described further on in Chapter 5, 'Impact of renewables and secondary heat sources'.

A single-head system operating alongside a conventional central heating system, as described above, is relatively cheap to install (in our experience around £1,500 supplied and fitted for a typical 4.5kW system and slightly more for a higher output system) and have a short payback time which can measured in years because of their low acquisition cost. With an expected lifespan of around 20 years, their £80 annual depreciation cost is modest compared to other renewables and to the amount of heat they produce.

One minor disadvantage of these systems is that whilst it is possible to wire the smaller units into a conventional fused spur on an existing electrical ring, the larger systems need to be wired directly to the house's electrical consumer unit and are, therefore, potentially more disruptive to install after the initial build. The same applies to air-to-water systems

A final brief cautionary tale about heat pumps: some companies quote very high costs to install these. At the time we installed our own first two systems, for the cost of diesel, a decent takeaway curry and a bed for the night, we shipped in our air conditioning engineer from Bolton in Lancashire some 150 miles away! Any owner quoted figures significantly over £1,500–£2,000 for a small (e.g. 4.5kW output) system should shop around as they are not intrinsically expensive either to buy or install. Larger systems do not cost that much more.

5. Other heat sources
- Biomass boilers

Biomass boilers are designed to replace fossil fuel-based central heating systems with carbon-neutral ones. They produce equivalent temperatures of hot water to a conventional gas or oil boiler and operate with a standard radiator system. They typically cost in the region of £30,000 and need space to house the large boiler and hopper system. Because they run continuously, they may (or may not) reduce the actual cost of heating compared to gas or oil and with a typical lifespan of around 20–30 years their depreciation costs are high.

- Multi-fuel and woodburning stoves

Although these are obviously often used as a main feature in a room, with the enormous attention paid to renewables it is easy to lose sight of the large effect they can have on reducing carbon footprint and the use of a fossil fuel central heating system.

When wood alone is burned they are carbon neutral compared to leaving the wood to decompose naturally, although many people will also add small amounts of smokeless coal or anthracite as a base to the fire. In environmental terms (and this also applies to biomass boilers) their benefit, in reality, is more than simply being carbon neutral. Unless the wood burned would otherwise be recycled for other purposes, all of the heat produced will effectively be a by-product of the carbon-neutral effect of burning wood as opposed to leaving it to decompose. Adding a gravity-fed system with radiators to a fire with a back boiler is cheap to install at the time of the original build, pays for itself rapidly in terms of the heat generated and can hugely reduce the amount of fossil fuel burned. The capital required to install, for example, two fires with back boilers with a potential heat output of around 30kW (if bought and installed cost-effectively) will amount to less than half of the cost of installing solar panels and around one-tenth of the cost of installing a wind turbine, yet will produce up to ten times the amount of energy as heat. With an existing chimney and using the owner's own tradespeople, it may not cost much more than £2,000 to buy and install a stove with a back boiler and additional radiators at the time of the initial build. Far more can obviously be spent on a stove if an owner wishes, although for the very cost-conscious there is a good market in nearly-new stoves, although UK DEFRA-approved (Ecodesign) versions are rarer.

Without a back boiler, most stoves will generate 5kW or more of direct heat and the corresponding cost arguments are also compelling.

Particulate emissions responsible for pollution remain an important issue with woodburning stoves even if they are carbon neutral. Ecodesign stoves, now mandatory for new fires manufactured in the UK, may reduce small particle emissions by over 80% compared to older stoves, and, particularly, open fires and when these are combined with burning 'dry' wood (moisture content ≤ 20%) emissions are reduced by a further 60% or so.

Using the various combinations and permutations of air-to-air source heat pumps and multi-fuel or woodburning stoves alongside a conventional central heating system can hugely reduce the use of fossil fuels. Although not an ideal environmental solution compared to purely renewable heat sources, they do provide the additional aesthetic attraction of offering the stoves as a feature point in a room.

Maintenance and running costs: impact of building solutions

This book does not consider maintenance or running costs other than heating in any detail, apart from describing a few things which can be carried out at the time of the original build which can have a large impact on future maintenance and running costs. In terms of maintenance, for example, pointing stone walls with a lime mortar can be frighteningly expensive and may require more regular repointing (not a cheap process) compared, if it is possible in terms of wall construction, to the option of adding a small amount of cement to the pointing mix for strength and durability. Similarly, renewables and also, in particular, other secondary heat sources (such as multi-fuel stoves), may be far easier and less disruptive to install at the time of the build than after it has been completed. Careful thought is needed as to how and where these will be used. It is, for example, relatively easy to add an air-to-air heat pump at almost any time, but adding additional radiators to a multi-fuel stove and back boiler system or a multi-head heat pump, may be disruptive. Owners should also take into account the likely use of the different rooms of the property during the day. This may seem self-evident, but what seems obvious at the start of a build may not actually be the case in practice once the build has been completed.

Overview

Several of the above sections interlink or overlap and the overview below is in approximate chronological order.

1. Grants may be available for some works on an old building, but these may be modest and carry strict conditions. Applying for these can also introduce significant delays, at the end of which it is possible that the owner will be told that no funds are actually available. It would be prudent to consider any grant as a windfall and not factor it into the budget.

2. VAT can vary greatly depending on the type of building, when it was last occupied and on the planned works. Unless the project is being managed entirely by an architect, advice from a tax specialist is extremely helpful to assess this in advance. Potential owners should research the VAT status of the work, if possible before buying the property. Single tradespeople who work mostly on a 'fit only' basis may keep themselves below the VAT threshold, which will reduce VAT liability on the labour if this is payable, although VAT may need to be paid on materials and depending on the building cannot necessarily be reclaimed. Again, a tax expert can advise.

3. The owner should ideally instruct a surveyor to assess the building, although in several of our projects we asked our builders to do this.

4. The owner will need to decide whether to instruct an architect and, if so, whether to instruct them to manage the whole process or only to complete some stages (such as producing plans). In particular, the owner will need to decide whether to ask an expert (typically an architect) to produce the detailed specifications or draw these up themselves or with a builder/tradespeople as the work proceeds.

This is not the only way of working, as surveyors, architectural draughtsmen and planning consultants can all also be instructed in different ways for different stages of a project.

5. The building contractor or tradespeople can then be chosen, if possible from personal recommendations, working either on a day rate or from quotations. If quotations are used, a slightly higher quotation from a personally recommended source is likely to be preferable to a lower quotation from an unknown source and it would be wise to check out the unknown source carefully and to read reviews where these are available: a bit risky if none are available. Day rate payments can work perfectly well and may be preferable for some more complex works, providing there is mutual trust between the builder/tradespeople and the owner.

6. It is more likely that changes will be made to the works during the build with old houses rather than barns, where the build options may encounter fewer unforeseen works. If the owner contracts the work on the basis of a quotation, then further quotations will be needed for these changes. Despite the potential disadvantage of not being an incentive to work quickly, working

on a day rate basis can also provide flexibility, enable the use of reclaimed materials and remove the 'honey with extras' possibility of high costs from a change in build specifications after an initial fixed quotation.

7. A tradesperson or contractor who insists on using their own materials may not be ideal for a project on an old building, particularly if the intention is to create an 'older' final appearance. This may prevent the owner from searching for either less expensive materials of equivalent quality or reclaimed materials.

8. Reclaimed materials can provide a better finish to the building and may (although not necessarily) be far less expensive to buy than new equivalents. They may, however, increase the labour fitting costs.

9. Ignoring their effect on carbon footprint and Government incentives, the economic viability of some of the available domestic renewables options is questionable and the money may be better spent on other renewables options. Ground-source and air-to-water heat pumps are not well suited to many old properties. Air-to-air heat pumps used alongside a conventional central heating system with or without multi-fuel stoves are arguably the best option for these properties: they greatly reduce carbon footprint, provide immediate heat, can be switched on and off at will, are inexpensive compared to other options and are easy to install.

As discussed in the individual projects in Chapter 4 (Page 73 onwards) we used standalone air-to-air heat pumps to heat individual rooms alongside conventional (oil-based) heating systems.

At the time we installed them, the water-based systems which deliver low-level heat were not well suited to older buildings. They also needed dedicated large radiators to deliver their heat. The possibilities for water-based heat pumps may change with the introduction of high-temperature pump systems similar to conventional central heating, although these systems are less efficient than conventional heat pumps and the savings made are more modest. It is unclear at present whether these can successfully replace a conventional heating system.

In retrospect, our choice of strategically located air-to-air type heat pumps combined with multi-fuel stoves operating alongside a conventional heating system is still our most cost-effective solution both in terms of installation and running costs. We can also switch these pumps on and off at will and only heat a room when it is in use (see Page 143).

The decision to install an air-to-air heat pump is straightforward and simply involves buying a system and employing an air conditioning engineer to install it. The only specialist advice needed is to calculate the power of the pump required (a standard 4.5kW output pump comfortably heats a typical 25m2 (or 50m3) room in an old property. They can also be installed during an initial build or be added later. Any more complex systems, whether air or water-based, are likely to need to be installed during the initial build if later disruption is to be avoided. They will also need specialist advice. Owners will also need to do their sums carefully to ensure that any decision really is cost-effective.

Chapter 3: Case studies – Background and descriptions

Excluding the early projects and Gin Gan described very briefly below, there are seven case studies. These were all very different in size and style, though there were some common works in three of the projects on the same site.

In chronological order the projects consisted of:
- The Mill
- The Garage
- The Farmhouse
- The Barn and its Gin Gan
- The Byre
- The Farmhouse Extension
- The Barn Conversion

The early projects do not merit any detailed description but they were useful in providing some initial, although limited, experience about the building process and problem solving.

The first house had been lived in by an elderly lady and little had been done to it over the years. However, unlike some of the houses in the terrace, it did at least have an indoor bathroom. Apart from installing a heating system, the remaining works were mostly remedial. Work was carried out using the same builder, based initially on quotations and subsequently on an hourly rate. There was limited scope to shave anything off the costs apart from my use of a single, non-VAT registered tradesperson; a nice but irascible chap with whom I had a very good relationship, which was probably just as well as he admitted to having a conviction for smashing someone's car windscreen following a disagreement.

The next three properties were Edwardian tenement flats in Glasgow. The work was similar to the upgrading of the Oxford house and involved installing a central heating system, sanding the original floorboards (something I did myself), stripping and waxing the painted doors and then full decoration.

The Gin Gan was a later project and the first carried out after we had moved to County Durham. It was a barn conversion, but one in which the building was purchased with planning permission and approved plans and converted under contract with the seller of the building. It therefore does not justify much more description as our only real involvement was in the choice of flooring and finishings.

When we moved to County Durham and discovered the beauty of North East England we decided to stay. It took us some time to find a place to buy as the stock of medium-sized houses in the county is quite small and most villages consist of small terraced houses. On the market at the time there were a number of new-build estates and a few cavernous 10-bedroom manor houses in the neighbouring dales belonging to former small estate owners. The latter were probably on the market because their annual running costs exceeded the gross domestic product of a small country!

However, there were also quite a few former farm buildings in the county, many of which were owned and sold off by the UK Coal Board (as it was at the time), bought by developers and converted into several living units. We bought one of these buildings located close to where we were renting; it was known as the Gin Gan as it included one. The building was essentially a rectangle with a north-facing gin gan protruding from the middle. It was part of a farm complex the builder was converting into seven residential units. The overall structure was solid and most of the building works were internal, with the exception of window openings for which the various permissions had been granted. All the masonry, first-fix joinery, plumbing and electrics, finishing joinery, groundworks and services were included in the building contract and we therefore had no specific input here. We did witness the work being carried out which, like the early projects, provided valuable experience.

As with the earlier upgrading projects some of the main decisions were around fixtures and fittings in kitchens and bathrooms. The style of these is obviously personal preference, and the means of finding and buying them were far more limited than today as the internet did not exist in the 1980s and internet search engines did not exist at the time of the Gin Gan build.

The Mill

This project is an example of the change of use of an old derelict building, but one with a sound fabric and the presence of both lower and a few upper floor window and door openings. In this project, we intended to create a modern finish yet preserve the main historical features of the original mill building including the mill workings.

Case studies – Background and descriptions

Photo 1 On the way to the Mill. Gateway to the real north: the Dornoch Firth and Kyle of Sutherland where sheep take over from humans along the A836, a circuitous route back to the northbound A9 over the Struie Hill. On my last trip to and from Inverness to the Struie turning, around 12 miles away, I counted eight red kites and countless buzzards.

A large number of visitors to Scotland do the standard thing of visiting Edinburgh, the mecca in most guidebooks (although being a Glaswegian I cannot understand why!), and often do not venture further north. Those who do go further afield even more rarely go beyond Inverness, apart from the few faithful who make the beautiful north west coast a regular venue. In early 1994 whilst we were living in Glasgow, we bought a derelict former Mill in Sutherland (not Sunderland as many of our acquaintances thought, probably because they had never heard of the former!). This was our first relatively large project. It also posed the challenge of being managed from a distance; Glasgow being a few hours' drive away. Little, however, did we know that we would be moving to Durham shortly afterwards.

The building was a late Victorian former mill built into a hill and therefore offering access on both levels which, the local farmer told us, was in use until the late 1950s. It was Scottish Grade C Listed and retained all of the milling equipment in the large lower area, the internal cast iron workings at one end and a large wooden fuel hopper at the other. The hopper still carried a pencilled date and signature marking the opening of the Mill late in the 19th century, a marvellous historical record. The smaller area contained a grain kiln and upper drying area. The building was relatively modest in size with a footprint of just under 90m^2. The kiln was a brick structure in the shape of an inverted square funnel occupying much of the smaller lower area leaving four narrow walkways around it. The top floor housed the upper mill workings with one remaining millstone at one end and the kiln drying area in the other, with a large open space between them. The mill race ran to the side of the building onto the mill wheel and drained down to a river. This had been diverted some time earlier by the estate which had owned the property.

The larger lower area had four window openings to one side, two of which were covered by banked up earth, and three on the other northern, and therefore darker, river-facing side. There were also three windows in the lower smaller space, one of which was also obscured by banked up earth. On the upper floor there were five window openings. The smaller space on the upper floor of the kiln contained only one window, although it also had three rooflights and at some stage had a cupola acting as a chimney. The Mill had a slate roof with a total, including the upper kiln area, of nine rooflights.

Although the build project appeared straightforward, it raised a number of problems as it was so remote.

Plan 1 shows the initial profile of the building.

In our opinion, for minimal impact on the building fabric and spaces, the best layout was relatively obvious. We planned to create a kitchen in the smaller lower area containing the kiln, a large open-plan sitting/dining area in the larger lower space and bedrooms with bathroom(s) on the upper floor, combined. A second reception room area in an upper entrance extension was also planned, as this was south-facing and warm on sunny days (contrary to what many may think this is not uncommon in the far north east of Scotland).

Architecturally, the main features on the lower floor were the mill workings and the large wooden fuel hopper, together with the open internal surface of the brick kiln once we had decided to cut it in half vertically. In the upper area, the main features were the upper part of the mill workings with the remaining millstone at one end and a star-shaped wooden structure supporting the funnel-shaped kiln roof at the other. The walls were coated with a lime render and whitewashed with no exposed stone. Highlighting these features within a modern restoration seemed a reasonable choice, did not raise any particular difficulties and needed only a simple building approach. We therefore decided to leave the appearance of the walls but dry-lined them and finish them in the same white colour.

The spaces on the lower floor would be largely untouched apart from splitting the kiln to form the kitchen space and removing a small, badly damaged wooden façade separating the mill workings from the larger area so that both of the main features at either end of the space were exposed.

On the upper floor, apart from the vaulted ceiling for the mill workings and the star-shaped arrangement of timbers supporting the funnel-shaped roof in the drying room, the addition of the missing cupola would be a bonus feature.

In terms of services, the Mill had an electricity supply, a mains water point was available 100m away at the end of the drive which did not cause any problems as a trench of that length is not difficult to dig and we owned the land, but had no sewage system.

The final intended floor area was approximately 210m² and the initial layout and final approved plans can be found in **Plans 2–6**.

The set of photographs which follow these (**Photos 2–14**) show some of the earlier images of the building.

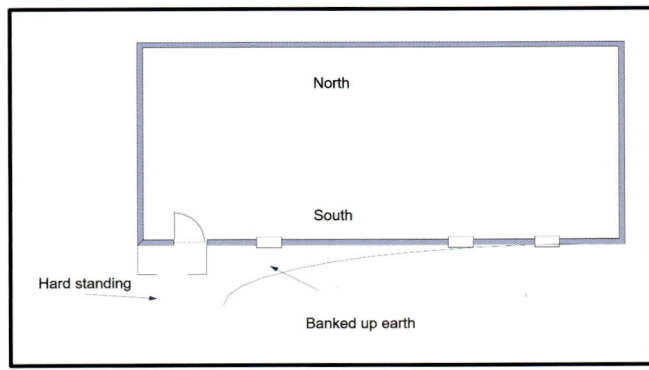

Plan 1 Mill. Initial outline plan of the Mill building (not to scale) with banked up earth completely obscuring one window and leaving only a lintel of the second exposed.

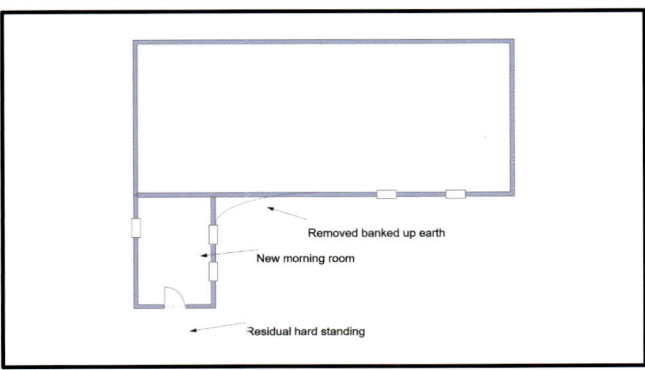

Plan 2 Mill. Final approved outline plan of the Mill building (not to scale) with part of the banked up earth removed showing the two lower floor windows and the new extension to the building where the original lean-to had been.

Old Buildings: Conversion and Restoration

Plan 3 Mill. Drawing (not to scale) showing the existing layout of the lower floor of the Mill building.

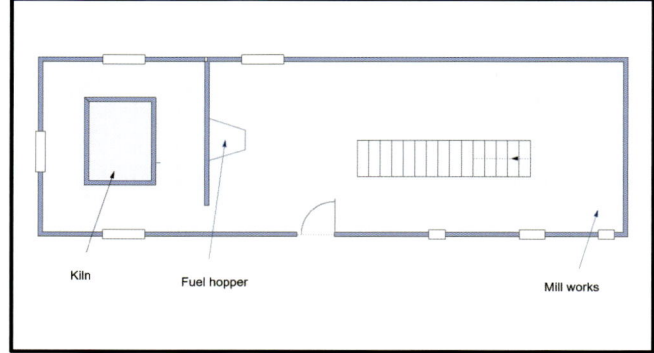

Plan 4 Mill. Drawing (not to scale) showing the final approved layout of the lower floor of the Mill building.

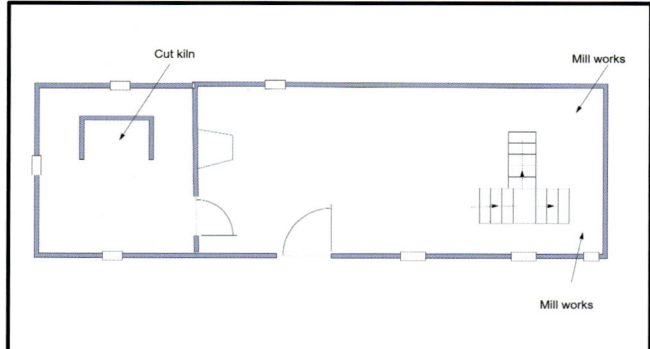

Plan 5 Mill. Drawing (not to scale) showing the initial layout of the upper floor of the Mill building. This was submitted along with the planning proposal to show the lean-to.

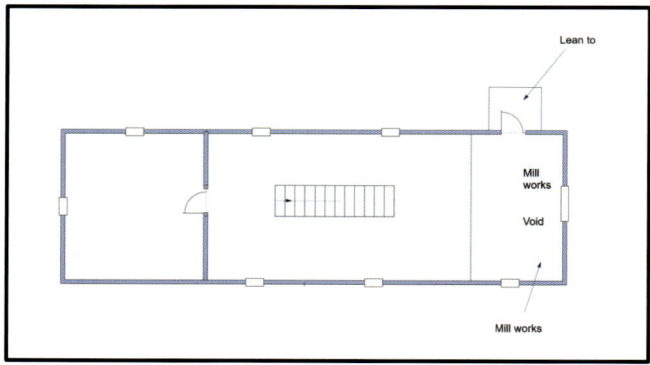

Plan 6 Mill. Drawing (not to scale) showing the final approved layout of the upper floor of the Mill building.

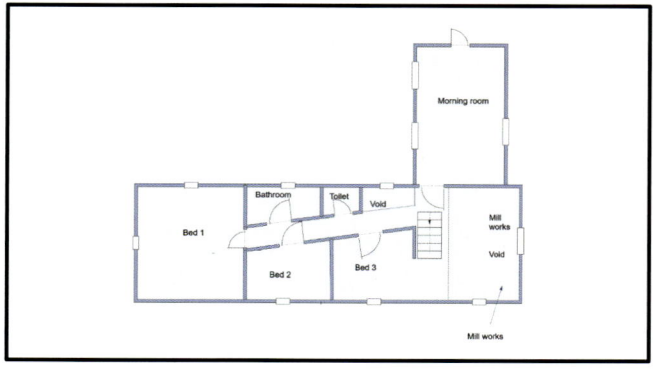

Case studies – Background and descriptions

Photo 2 Mill. Front of the original building.

Photo 3 Mill. After removal of earth covering the windows (and the tree).

Photo 4 Mill. The Mill wheel required only a tidy and a spray with a proprietary preservative.

Photo 5 Mill. Appearance of internal wall with roofing timbers.

Photo 6 Mill. Detail of the rear of the Mill building.

Photo 7 Mill. Upper floor, doors to smaller grain drying area: iron rails support the perforated cast iron tiles (see Photos 25–29).

Photo 9 Mill. Larger part of the upper floor facing away from the smaller area, looking towards an east window.

Photo 8 Mill. Upper floor. Opposite direction. Rather daunting.

Photo 10 Mill. Minor repair and cleaning works restored the mill workings.

Photo 11 Mill. Door to the lower smaller area to kiln with a surrounding 'corridor'.

Photo 12 Mill. Kiln cut and part-finished second 'corridor' area used to house the boiler.

Photo 13 Mill. Original mill workings obscured by a damaged façade.

Photo 15 Mill. A tranquil spot to enjoy a glass of Chablis and the sounds of running water and the occasional screech of a buzzard overhead.

Photo 14 Mill. Mill workings after removing wooden façade and cleaning.

Photo 16 Mill. A picture-postcard view at the end of the garden. Location, location etc.

The Garage

This is an example of a simple conversion of a modern building from a single open space into a small dwelling on the same site as the Gin Gan. So, whilst not a 'period building conversion' it is included as it illustrates some of the basic issues and solutions available for conversion of any building. This was the first project we managed ourselves and would have been ideal to carry before the Mill as it gave us first-hand experience of project management and some of the decisions we would need to make. Unfortunately it came later.

The details of the Garage conversion project are probably

irrelevant to anyone with some building experience but are included for those who do not; giving some examples of cost-containment and the simple techniques used. The plan was to convert the garage into a space that could be used as a nanny flat/annexe. The garage was a rectangular building, 7m long and 5m wide, and was a dry, insulated shell constructed with a stone exterior and blockwork interior. During the initial Gin Gan project one of the few building decisions we had made was to add both a damp proof membrane to the floor and insulation to the walls of the garage. Until then the building had acted as a glorified storage room (the adage that one expands to fill the space available!). As this was a simple conversion within a completed building, we did not have to make any architectural or complex building decisions. We designed it ourselves as a self-contained one bedroom building.

The final intended floor area, lower and upper floors combined, was 50m^2 and the final approved plans of the building are shown in **Plans 7 and 8**.

The Farmhouse, the Barn and the Byre: overview

Although the plans and initial works relate to one set of buildings, these made up three separate projects which we carried out as individual builds in three separate phases and on

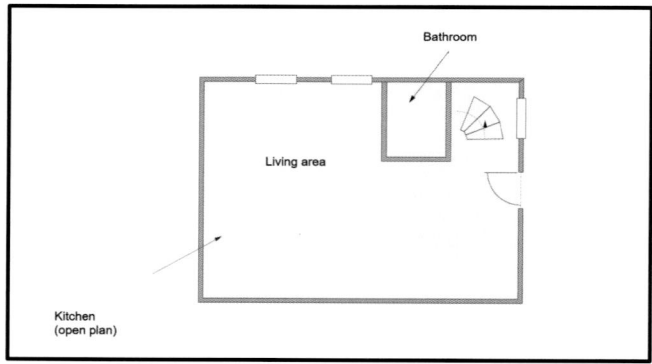

Plan 7 Garage. Drawing (not to scale) showing the final approved lower floor of the building.

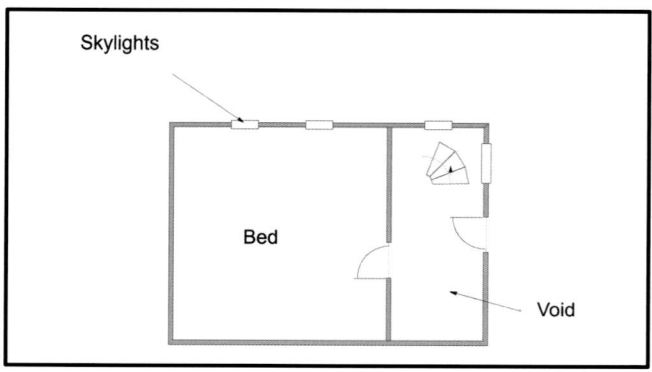

Plan 8 Garage. Drawing (not to scale) showing the final approved upper floor of the building.

three quite different building types. These three projects (dealt with under separate sections below) involved:

1. restoration of a period farmhouse,
2. conversion of a large barn and gin gan into two open living spaces keeping all of the original features, and
3. conversion of a former cow byre into a one bedroom dwelling, modern in part but keeping a number of original features.

The whole property has now been our resting place for over 20 years and we intended it to be our last, so we wanted to get the conversions and restoration right. The whole farm complex of buildings carried an English Grade II listing because of the Barn, which was thought to be a northern version of a longhouse. This consisted of two areas: a larger part (shown as Barn area 1 on the plans) and a smaller part (shown as Barn area 2 on the plans). Longhouses, mostly found in the south of England, consist of a roofed barn the main part of which would provide shelter for livestock with a smaller area that would have been a shepherd's kitchen, a wooden staircase would lead to a small upper floor which would have acted as the shepherd's bedroom. In our case the wooden staircase was badly damaged and had a wood panelled balustrade. The Barn attached to the Farmhouse led directly on its opposite end to a former cow byre (the Byre), creating a courtyard between the Farmhouse, Barn, Byre and a further small block of outbuildings. The overall layout of the whole farm complex is shown in **Plan 9** and the overall layout of the lower and upper floors of the farm buildings are shown in **Plans 10 and 11**.

Photos 18 and 19 show before and after aerial views of the whole complex of buildings and **Photos 20–25** show a number of views of the outside of the overall farm complex before the main works began.

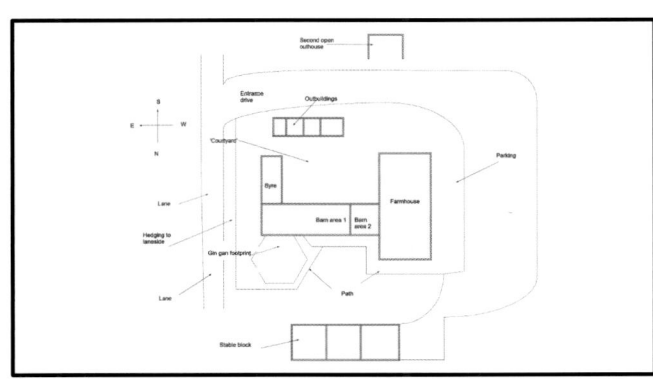

Plan 9 Farm complex. Drawing (not to scale) showing an overview of the farm buildings with the arrangement of all of the buildings. The larger and smaller Barn areas are shown as Barn area 1 and Barn area 2.

Old Buildings: Conversion and Restoration

Plan 10 Farm complex. Drawing (not to scale) showing the existing layout of the lower floor of the farm buildings. The larger and smaller Barn areas are shown as Barn area 1 and Barn area 2.

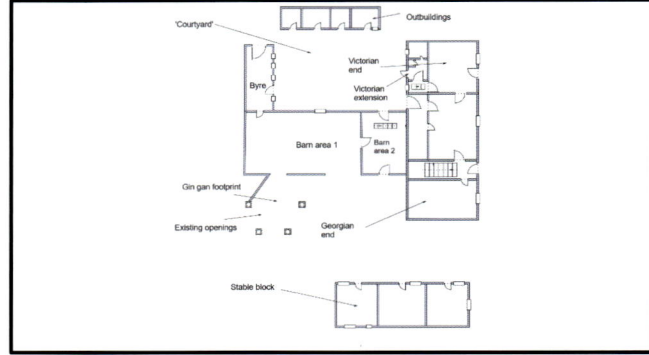

Plan 11 Farm complex. Drawing (not to scale) showing the final approved layout of the lower floor of the farm. buildings. The larger and smaller Barn areas are shown as Barn area 1 and Barn area 2.

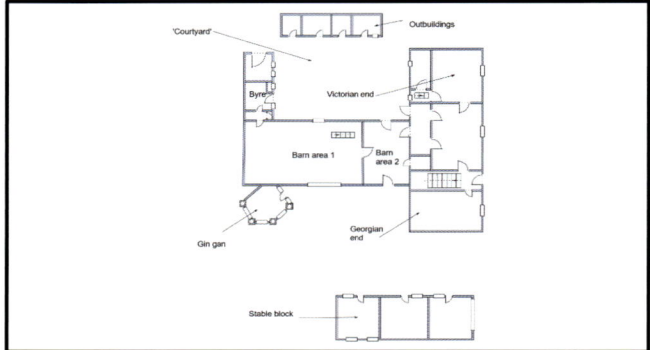

Photo 17 The North East has a host of 'don't miss' venues from sea to dales. Tynemouth market is great for browsing and street food.

Photo 18 Farm. The farm buildings had to be stripped out internally, but the building fabric was sound apart from the chimneys (here).

Photo 19 Farm. Aerial photograph of the farm (1964) showing, amongst others, the additional structure in the courtyard against the barn.

Photo 20 Farm. Recent aerial photograph of the farm from a slightly different angle.

Case studies – Background and descriptions

Photo 21 Farm. Early view of the 'Victorian' end of Farmhouse.

Photo 22 Farm. Early view of the back of the barn showing the gin gan.

Photo 23 Farm. Early view of Victorian extension showing three access doors.

Photo 24 Farm. The Marley tiles are not out of place against the Barn pan tiles.

Photo 25 Farm. View to the Byre. Building sites make great playgrounds.

The Farmhouse

This is an example of the restoration of a derelict period house, in this case a farmhouse, which had not been lived in for 50 years and needed to be entirely stripped back to its internal wall structure and then put together again as a period house. The layout was similar to many Durham farmhouse buildings in that it consisted of a four-roomed Georgian-looking house, which would have been occupied by a farmer, attached directly to a two-roomed Victorian cottage that would have been used by a farmworker. It had a total floor area of around 210m² over two floors. At some time in the building's history an opening had been created between the two parts. What was unusual was that the Farmhouse had a narrow, 2m-wide extension, probably built in Victorian times, that stretched much of the length of the Farmhouse attaching it directly to a barn although there was no direct access between them; the barn was probably built probably in the 18th century. The Farmhouse building was roofed with 20th century Marley concrete tiles. Our intention was to recreate the appearance of the original farmhouse.

The house was totally derelict apart from one room which had been occupied a few years previously, quite how I do not know given its monastic state. There were some basic facilities including a sink and water supply, the latter was quietly dripping from under the sink into the neighbouring wall, and a single electricity cable leading to a socket and light bulb. There was a floor of sorts, but only one the brave would tread on. The internal dividing walls were covered with lath and horsehair plaster, most of which was falling off, and the ceiling plasterboard was falling apart. There were a few remaining scraps of wallpaper on one stairwell which a local businessman we knew recognised from the days he had lived on the farm as a boy some 50 years previously. Between the Farmhouse and adjoining cottage there were half a dozen cast iron fireplaces: a rather magnificent Victorian range in the lower room in the Victorian cottage, a frankly ugly 20th century stone fireplace in the lower middle room of the Farmhouse which had become seriously outdated and various cast iron Victorian fireplaces in other rooms in both the Victorian and Georgian parts.

We could have used the Farmhouse rooms for any purpose although the central room on the lower floor leading to the original Victorian cottage on one side and the remaining rooms of Farmhouse on the other had four door openings and therefore logically was most suitable as a sitting room in the centre of the house. We opted to use the remaining five rooms as bedrooms. We decided to create three bathrooms, one from an existing bathroom in the Victorian extension, one at the other end of the same extension and one from a space 'stolen' from the upper room of the Victorian part of the building by moving a stud partition wall, leaving easy access to a bathroom from all rooms. We also planned to 'steal' the smaller area of the Barn to act as a kitchen; it had three existing doors, two external doors on either side and a door into the larger part of the Barn and we planned to add a new door into the main Farmhouse. This area also included the upper floor 'shepherd's bedroom mezzanine' which we intended to use as an office/study. Together these upper and lower floors of the smaller Barn area added 50m^2 to the Farmhouse.

The Farmhouse windows were rotten as were many of the four external doors. There were a few viable internal doors in the Victorian end, of panel construction and typical of the era. Although the Farmhouse had been reroofed with Marley concrete tiles, the original pantiles were still present on the roofs of the Barn and gin gan, Byre and the small outbuildings block in the courtyard. The stable block on the north side of the property had been built in the 19th century by a well-known North East England farm building company and still bore the company's markings on sinks in the building. This set of outbuildings had a sound slate roof. In The Boys' opinion the whole building fabric appeared sound, it had electricity and a water supply and it had somewhere for sewage to go. The roof also appeared sound, although much of the guttering was wooden and had rotted away in many places.

In terms of services, the water and electricity supplies were rudimentary. Electricity came directly from a transformer some 50 yards from the property but still on our land, so did not raise any problems although we did bury the overhead cable during the initial groundworks. The water supply came from an existing piped mains system located 1km away, so we did (and still do) take on the risk that we could incur significant costs should the long pipe leak or burst at any stage. This caused our solicitor some indigestion, although realistically we had no option other than to accept the situation and accept the ongoing risk. In reality over 20 years we have not had any significant problems.

Maybe we have just been lucky, but some risk-taking with property ownership is almost inevitable. The Farmhouse had a wonderful ceramic sewage system, a tribute to Victorian engineering which would not have looked out of place in a Brunel exhibition, together with a brick septic tank and soakaway which may have started life as a cesspit and at some time acquired its soakaway. Although this looked rather basic compared to the sleek plastic things available nowadays, it was large, sound and perfectly functional. Services were not therefore a major issue.

Architecturally, the features which were visible initially in the Farmhouse included the exposed top and underside of the flooring in the middle room, the Victorian range and the fireplaces, and the vaulted ceiling with its beams in the original smaller part of the Barn (stolen to form the kitchen for the Farmhouse). The final intended floor area of the Farmhouse, including the 'stolen' smaller part of the Barn and its mezzanine, was approximately 260m². The initial and final approved Farmhouse plans for the lower floor are included in Plans 10 and 11 shown earlier. The upper floor of the Farmhouse was unchanged apart from relocating a modern stud partition wall to create a bathroom.

The Barn

This is an example of a project carried out on a structurally intact barn, which we planned to use as a large open plan 'reception' space. The later Barn Conversion (see Pages 69–71) description shows how we converted this space to create a separate dwelling (a true barn conversion). Our intention in this project was to leave the area open as a barn, as close as possible to its original appearance.

Because we intended to add this area to the Farmhouse accommodation created in the previous project no room partitioning was required and only minimal plumbing (for central heating). The remaining part of the Barn, after 'stealing' the smaller part for the Farmhouse kitchen, was approximately 16m long by 5m wide. It also included a gin gan, which was hexagonal in shape and approximately 7m at its widest. This larger Barn and its connecting gin gan would create additional reception space. The Barn had no upper floor apart from the mezzanine above the smaller Barn area which was incorporated into the Farmhouse.

All of the internal and external walls in the Barn and the pillars in the gin gan were exposed. The roofing structures of the Barn were also exposed, as was the gin gan roof together with its star-shaped wooden supporting structure and heavy beams to support the weight of the (missing) millstones. In the smaller part of the Barn there were badly damaged remnants of an old brick bread oven built in front of what appeared to be a large old brick fireplace. The Barn had three external openings, one large and two relatively small. It had no damp course and no services were present to this part of the building.

The features we wanted to keep in the Barn were the roof purlins and A-frames in the Barn itself and the roofing timbers and supporting structures for the missing millstone in the gin gan. We decided simply to leave the barn as an open plan space but to create an upper mezzanine, mirroring the mezzanine above the kitchen.

We had opted for no partitioning or separation other than the door from the barn into the gin gan and simply intended to use the different parts of the barn for different purposes. We planned to have a small casual 'snug' at the Farmhouse end below the new mezzanine, a central dining area and a larger more formal seating area at the opposite end to the Farmhouse and in the gin gan. The final intended floor area of the Barn including the new mezzanine was approximately 160m^2.

Case studies – Background and descriptions

The initial and final approved layout of the Barn area with its gin gan are shown earlier in **Plans 10 and 11 (Page 62)** and the initial and final approved layout of the upper floors of the Barn are shown in **Plans 12 and 13**.

The Byre

Like the Garage project this is an example of how we used a relatively modest space to create an individual dwelling, but in this case in an old building with the basic structure (including rotten windows) already present. We intended to create a modern appearance yet preserve some of the historical features of the building.

The Byre building was around 14m x 5m in size and was originally a single concrete-floored space with six concrete animal stall separators and attractive, but not especially old, roof purlins. We originally intended this to be a one bedroom apartment in the new building, like the previous nanny annexe created in the Garage. It has since become a let property and income source.

As mentioned the Byre had a vaulted roof with exposed roofing structures. It had two external doors, one at the opposite end from the Barn and one opening into the courtyard. Although the building fabric was

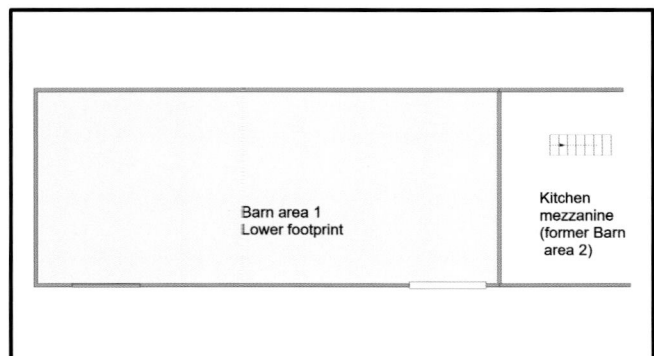

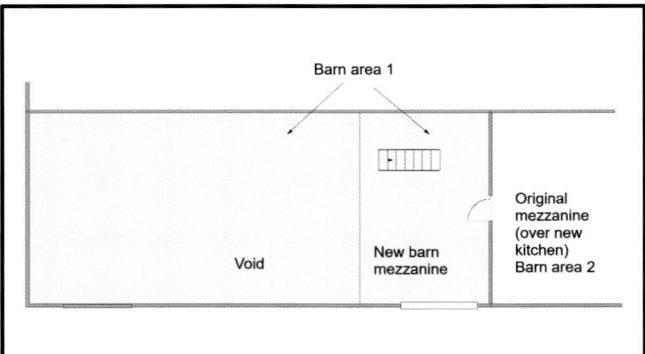

Plan 12 Barn. Drawing (not to scale) showing the existing layout of the upper Barn areas (gin gan not shown as it had no upper area). There is a single mezzanine over about half of the area of the smaller part of the Barn became the Farmhouse kitchen. The larger and smaller Barn areas are shown as Barn area 1 and Barn area 2.

Plan 13 Barn. Drawing (not to scale) showing the final approved layout of the Barn with a second mezzanine created above the larger Barn area accessed by the new relocated staircase. The larger and smaller Barn areas are shown as Barn area 1 and Barn area 2.

sound, the roof was in poor condition and appeared to need complete reroofing. Like the Barn, the Byre had no services. The architectural features potentially worth preserving were the roof purlins, wooden A-frames and some or all of the animal stall dividers; there was also potential to expose some walls.

Like all of the projects, we intended to follow the same principle of keeping the appearance as simple as possible, again to minimise any visible intrusion clashing with the original appearance. We did not intend to build any additional space beyond the original footprint of the building and the intended final floor area was approximately 70m².

The initial and final layouts of the Byre are shown in more detail in **Plans 14 and 15**.

The Farmhouse Extension

This was a very simple project. It is an example of a low-cost, sympathetic, modern extension to a period building (to match an addition present on the old aerial photograph shown on Photo 18, add page reference). The build was not complex and it is included because it gives examples of some decisions we made about its structure and our use of

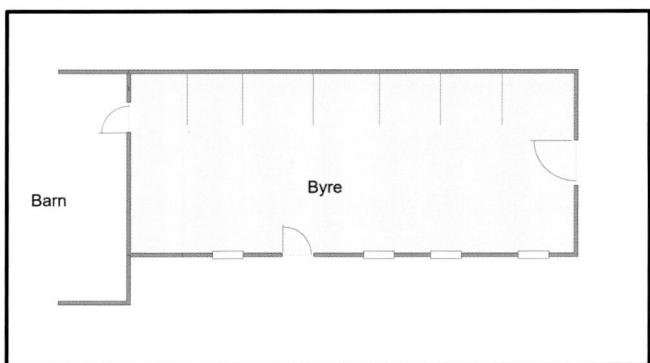

Plan 14 Byre. Drawing (not to scale) showing the initial layout of the Byre.

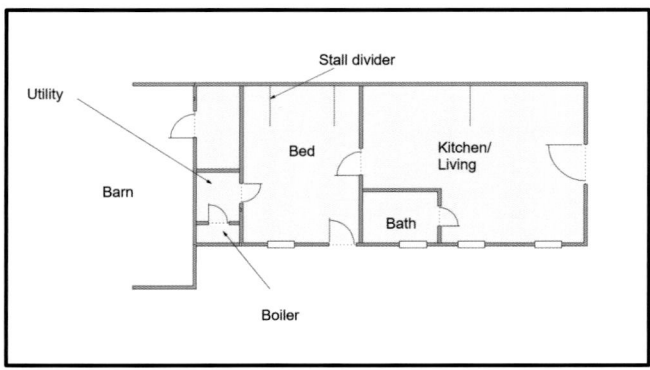

Plan 15 Byre. Drawing (not to scale) showing the final approved layout of the Byre.

Case studies – Background and descriptions

renewables (in this case an air-to-air heat pump), both to simplify the build and to reduce our consumption of oil in the Farmhouse as a whole.

The extension was an uncomplicated addition to the Farmhouse, effectively a glorified conservatory, but designed to act as an everyday morning room. Evidence of the previous addition was required to support our planning application and some imagination to replicate the previous lean-to structure which we discovered from the early aerial photograph (see Photo 20. Page 62). We tried to match the previous structure as closely as possible and to remain sympathetic to the overall appearance of the other buildings. No conversion or restoration was required as it was a new build.

The building solutions and materials used helped us both to:

- recreate the original appearance of the lean-to, at the same time significantly reduce building costs, and

- reduce running costs in the whole Farmhouse using an air-source heat pump installed in a heavily used room.

The final planned floor area was approximately 20m² and the position of the Farmhouse Extension within the farm complex is shown in **Plan 16**.

The Barn Conversion

This is possibly the closest example to a barn conversion in the conventional sense of the term, i.e. a dwelling created from a barn including kitchen, bathroom and bedrooms; although in this case we had carried out the basic works at the time of the original Barn project 16 years previously. Our intention was to create a modern living space but retain the original vaulted ceilings and exposed walls as in the original Barn project.

Our major challenge was to create bedrooms and bathroom spaces in the original open Barn (which we increased by approximately 25m² from its original size by adding a further

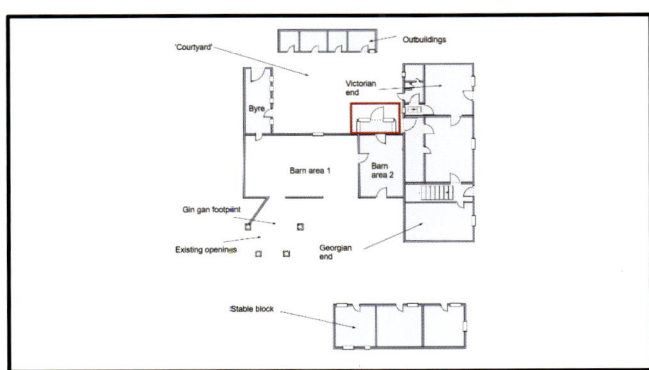

Plan 16 Farmhouse Extension. The position of the Farmhouse Extension (boxed in red) showing where it sits against the original farm layout, in almost exactly the same position as in the early aerial photograph (Photo 18).

69

mezzanine), without losing the appearance of a barn.

The initial and final approved layouts of the lower and upper floors of the Barn Conversion are shown in **Plans 17–20**.

The final approved planned floor area including the new mezzanine was approximately 190m².

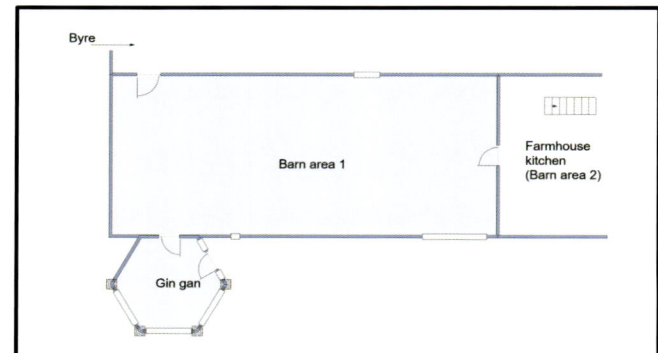

Plan 17 Barn. Drawing (not to scale) showing the initial layout of the lower floor of the Barn Conversion.

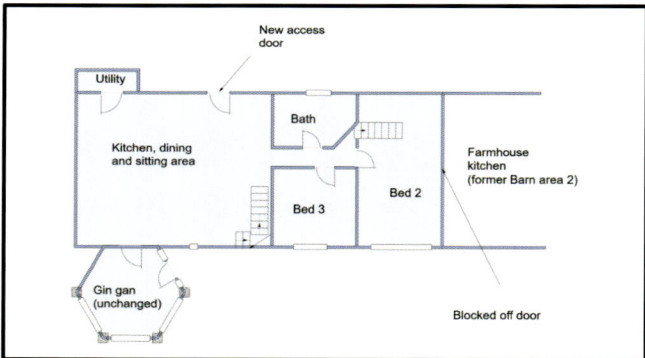

Plan 18 Barn. Drawing (not to scale) showing the final approved layout of the lower floor of the Barn Conversion.

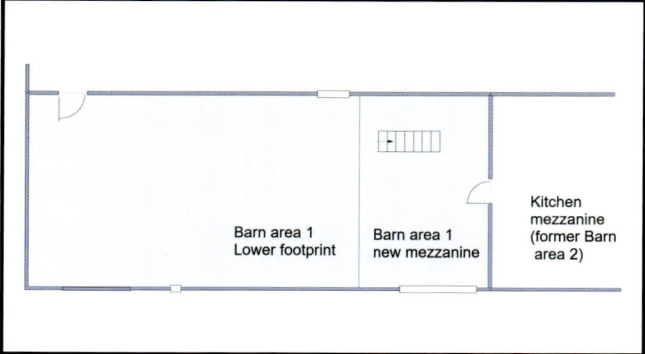

Plan 19 Barn. Drawing (not to scale) showing the existing layout of the upper floor (originally Barn area 1) of the Barn Conversion.

Case studies – Background and descriptions

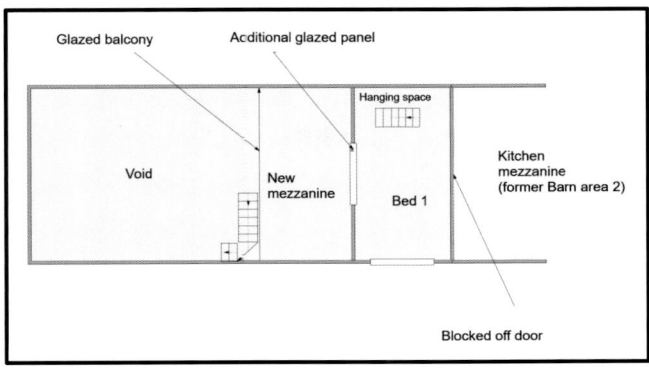

Plan 20 Barn. Drawing (not to scale) showing the final approved layout of the upper floor (originally Barn area 1) of the Barn Conversion.

Period building conversion and restoration

Chapter 4: Case studies – the builds in practice

This chapter illustrates our decisions made on each project, why we opted for these and the effects they had on costs and both immediate and later results. Although several of these projects were on the same site, we carried out the builds separately and the issues they raised were different. In some situations, the decisions taken contradicted the current advice given in the media and in other publications. All the main projects, apart from the Farmhouse Extension and the Barn Conversion, involved starting with a shell of a building, but there were differences in approach, some of which related to one, the Mill project, which we managed from 300 miles away, compared to on-site projects for the farm complex. Other differences in approach related to the nature of the existing buildings themselves. These experiences are obviously specific to our projects, although most should be transferable and relevant to others.

The projects are broken down into the sections listed below, grouping together some of the component parts described previously. I have grouped some of these sections in a few of the projects together as each is very short. Each section gives an overview of the main decisions which influenced either cost or the final result. I have discussed grants in Chapters 2 and 5 and these are not considered further in this chapter.

Component parts of a build:
- Building design and planning
- Build specifications
- Finding builders
- The build itself
  - Initial groundworks and services
  - Masonry works, roofing and guttering
  - First-fix joinery
  - Plumbing, heating, electrics, renewables and secondary heat sources
  - Finishing joinery, other finishing works, decoration and finishing groundworks

The Garage

As described in the previous chapter, this simple conversion was carried out after the Mill. I have described it first in this chapter however, as it provides a description of the basic component parts of a conversion as Gillian and I experienced them. Beginning as we did with the Mill and the challenges we were faced with, this first project was rather a baptism of fire compared to the second Garage conversion, which was neither difficult nor complicated. The description may be useful as we organised the works ourselves, with no expert input into the design or drawing up of specifications other than from our tradespeople and we also managed the stages of the project ourselves. The only external input was from an architectural draughtsman to produce the plans and navigate the planning process. It therefore illustrates some of the simpler decisions we were faced with, the decisions we took and how we shaved a small amount of money off the total potential costs.

Building design, planning and build specifications

We used an architectural draughtsman to create the plans from our own drawings, together with the elevations required for planning permission and building control, as the overall project was simple. This cost a couple of hundred pounds. Whilst someone with minimal experience might have used the services of an architect, we felt confident in designing the interior ourselves and defining the detailed specifications with our builders. The only changes to the building fabric involved creating window openings on the lower floor and installing two rooflights on the upper floor. Otherwise, the building fabric remained unchanged. We did not need a survey and did not plan to make any additions to the building.

The Garage conversion consisted of an open-plan lower floor with a kitchen area at one end and an entrance door at the other, with a small entrance area containing a reclaimed spiral staircase. A small upper hall led through a door in a stud partition wall into a bedroom, which was relatively modest in size as it was limited by the usable height dictated by the eves of the roof. We had no difficulty in obtaining planning permission, primarily because of the minimal external changes to the building. Plans 7 and 8 (see Page 60) illustrate how we used the building.

Finding builders

We were fortunate in that we did not have to look far for our builders, as we had met them during the original

Gin Gan build and had already worked with them on a small stable block built in an adjoining paddock. We were comfortable with their knowledge of the necessary specifications in this build and having worked with them on a day rate basis previously, we felt that this was a reasonable approach to work with them again. They have proved to be invaluable, in terms of their building skills and their knowledge of both building control requirements and technical options for the builds.

The build itself

- Initial groundworks and services

We did not need to carry out any groundworks except to dig the necessary trenches for sewage, surface water drainage and electricity. All were available in the immediate vicinity.

- Masonry works, roofing and guttering

This work simply involved creating openings for the windows on two sides of the building, looking away from the neighbouring part of the Gin Gan, and installing rooflights. Conservation rooflights were not needed (fortunately) and the roof itself was new and sound.

- First-fix joinery

This involved dry-lining and further insulating the walls, creating a bathroom with stud wall partitions, fixing joists to support the upper floor, installing the spiral staircase, flooring and finally insulating and plasterboarding the ceilings, none of which merit any detailed description. We did, finally, succeed in using a staircase which had travelled around the country with us: this was installed in the entrance hall giving access to the upper hall and bedroom. This was doubly successful as firstly a conventional staircase would have turned the design around and caused difficulties with the layout of the building, and secondly the space in which it was installed was not of any practical use. The overall cost of the staircase was probably about half that of buying and fitting this new, once the additional labour had been factored in. The external door (expensive to buy new) was reclaimed and was part-glazed, which increased natural light.

- Plumbing, heating, electrics, renewables and other secondary heat sources

Plumbing, heating (electric in light of the cost of running a separate trench for an oil supply for a central heating system for such a small space) and electrics were all straightforward; these were run below the floorboards and, where needed, in the stud partition walls. Domestic renewables were in their relative infancy at the time and we did not install any. We also did not attempt to install an open fire or stove.

- Finishing joinery, other finishing works, decoration and finishing groundworks

These were uncomplicated. We left the garage open-plan on both lower and upper floors, apart from the stud partition walls and doors (ledge and brace, which we made) to the bedroom and bathroom. The only other finishing works needed were skirting boards, cupboards built into the eaves, and other assorted minor works. Similarly, we used simple, modern, reclaimed kitchen units and bathroom fittings and we did not need any finishing groundworks. We kept the internal decoration simple, with emulsion paint and externally we added a new front wooden façade and the reclaimed door.

There were few particular building issues with this project.

The Mill

This section describes how we converted a derelict property into a dwelling with a modern interior, but retaining the main features in the building.

Building design and planning

Although we felt that much of the design was self-evident, we instructed an architect specialising in old buildings to design the building and create the necessary plans. Our vision was that the lower smaller area at one end of the building would act as a kitchen, with an open-plan area in the large lower area, rather than try to partition this into separate rooms. We planned to create bedrooms and a bathroom plus an additional toilet on the upper floor, using a diagonal corridor from the entrance area to the kiln drying/malting area. We also thought that we could create some additional space at the upper entrance to the building, where the lean-to had been located, in order to provide both a large entrance area and additional reception space. We felt that the existing window openings were sufficient to provide reasonable light to most of the building, except for the drying/malting area of the kiln on the upper floor where we needed to create a number of new openings.

The architect turned our ideas into a plan and offered a couple of useful additions. He proposed a rather grand steel staircase from the upper entrance area of the building to the lower floor and a void at the end of the small angled area next to the corridor to provide additional natural light to the lower floor. Our only other possible use of this small area would have been for additional storage space and although this might have been useful, we agreed with the architect that the visual effect of the void and the additional light it provided would be a better use of the space.

We resolved the question of creating two windows in the upper kiln end of the building in principle through discussion with the planning department and conservation planning officer, who were supportive. They were also sympathetic to adding the extension where the entrance lean-to had been located. The only planned internal structural change we wanted was to cut the kiln to make the space in the smaller area of the lower floor usable. The conservation planning officer raised no objections to this, provided that we left the cut surface and interior of the brick kiln exposed.

Build specifications

We began by instructing a surveyor who confirmed that other than what was glaringly obvious (the need for a damp proof course, woodworm treatment and replacement of windows etc.), the overall building and roof were sound. The survey provided sufficient background information to reassure us that the project was viable. The report, however, was peppered with the usual surveyors' caveats in terms of inspection of inaccessible areas etc., and really only gave us a very general guide about the state of the building. The architect's work was split into two conventional stages:

1. to prepare and submit the plans for approval, define the building works specifications and put the works out to tender, and
2. to supervise the build itself.

We found the benefit of our architect became more limited after the plan submission process in the Mill.

The specifications our architect provided were to a gold standard but included a number of works, which were extremely complex. One example was the approach to the wall separating the large and smaller lower spaces in the building which appeared, on first inspection, to be dry. The architect's proposed solution however was to excavate to 1m below ground level, tank the exposed area and then backfill. This would have required either large machinery or an enormous amount of manual labour to break through the existing concrete, likely, in the opinion of our surveyor, to be significantly more than the 60cm height difference between the larger and smaller lower spaces in the building. Furthermore, in order for any heavy plant to access this wall, the door opening on the lower floor would have needed to be widened, raised and then made good – at considerable additional cost. A second attractive, but extremely expensive, proposal was to fabricate the staircase in steel with hardwood treads, using tensioned stainless steel wires and turnbuckles instead of wooden spindles on the sides of the staircase.

The tendering process was broken down into around 30 sections. With only a limited number of local companies and tradespeople in the far north of Scotland, we received few tenders for the different works. When we took over the tendering process many of the tradespeople or companies that subsequently carried out the work told us that when they received the original specifications they had found them too complicated to quote for the work. What our architect had provided, therefore, was a gold-plated solution for the building which was very thorough, but which few people felt they could tender for.

We then instructed a firm of surveyors for advice and agreed a fixed fee for this service. In his first five minutes in the building the surveyor put a damp meter on the dividing wall between the two lower areas and advised that the wall in question, which the architect had proposed to tank, was completely dry and that tanking was almost probably not necessary. This proved to be the case over the 20 years we owned the building.

With the help of the surveyor, we reduced the initial masonry works from around £50,000 based on the original tender from a contractor to £10,000 in today's costs. After this, we instructed the same surveyor to take over the supervision and make occasional staged visits to the build. Building control was, as usual, involved both in the initial plan submission process and at stages of the build. These experiences were broadly positive.

Finding builders

The tendering process identified a number of contractors of varying sizes although we did not know them. When we took over after this process, we set out to supplement these and find individual tradespeople to work with.

The build itself

Most of the Mill project was ultimately carried out by individual tradespeople (stonemason/builder, electrician, plumber, joiner etc.). This was a rather awesome project, not only because it was the first complete renovation project we had carried out but also because of the prospect of running it from 300 miles away. In particular, after the initial masonry stage our original builder later had difficulty delivering on his later quotations for the further works, beginning with the first-fix joinery, which caused us some problems as we were living so far away. The staged visits we had agreed with the surveyor were not regular enough to pick up potential problems at an early stage.

- Initial groundworks and services

The major groundworks required, other than those to install services, was to excavate the slope which covered the lower windows on one side of the building in order to expose the windows. In this case the decision was straightforward as, remarkably, we had a small and reliable company based about two minutes down the road, which had a good reputation. We did not, therefore, go out to tender but asked simply for a quotation, which appeared perfectly reasonable. Ultimately, the groundworks were neither complicated nor expensive. It is impressive to watch the operators of these large earthmoving machines, half the size of an articulated lorry, who can manoeuvre their buckets with sufficient dexterity to lift a stone the size of a small potato without actually touching the wall of the building itself. In terms of services, we needed to dig a trench along the driveway to the main approach lane in order to access the mains water supply. The other service groundworks needed were to install a septic tank and soakaway, dig a trench to lay the sewage pipe from the building and install a new septic tank and bury the mains electric cable (not essential but done anyway for appearance). None of these raised any significant problems.

- Masonry works, roofing and guttering

The masonry works inside the Mill were uncomplicated and consisted of opening the two window holes in the upper kiln wall, cutting the kiln vertically in half to create the kitchen space and making good the cut edge. We also needed to lay a damp proof membrane and we tanked some areas of wall which were below the outside ground level. Tanking was done with bitumen in those days; good modern tanking renders are now available and bitumen is not. Whilst many experts will say that walls constructed with lime mortar should not be tanked, the walls below ground level were palpably wet, the walls were filled internally with rubble and we felt would behave more like a cavity wall than a solid one. We did not have any later problems other than when the neighbouring estate had worked on the drains at the top of the driveway and heavy rain resulted in a stream of water running down the slope outside the new morning room extension and through the foundations underneath the tanked wall. This then flooded the kitchen area, which was below the outside ground level. This was not therefore caused by the tanking. Although the estate remedied their works, we added to our initial groundworks and laid a simple metal grid flood drain, at the end of the concrete hardstanding on the driveway, which emptied into the Mill race. It is therefore well worth thinking laterally about any potential for flooding, apart from the obvious, as installing a flood drain costs pennies, particularly if a digger is already on site. We should perhaps have given more thought to the possibility of downward runoff from a hill about 1km away. Water does ultimately need to go somewhere even if it comes from far away. There is realistically little that can be done to prevent water from soaking into the depths of old foundations and potentially rising up and causing flooding. The only way of preventing this is to avoid the water getting there in the first place.

The door openings in the lower part of the Mill were higher than standard door height and it was therefore straightforward to lay a concrete screed and a damp proof membrane (DPM) over this before fixing battens for the flooring. The internal wall coverings were sound and very little needed to be removed in order to install dry-lining.

The next part of the masonry work in the Mill was to construct the extension room on the south end of the building where the original ramshackle lean-to entrance had been. This should have been built of stone to match the building although, in our absence, our builder constructed a blockwork cavity wall with stones inserted at places along the outside wall for appearance. This meant that rather than being able to finish the extension walls with a pointing mix, we used a form of pebbledash, leaving the few stones exposed. This was not ideal and certainly not our original intention. Fortunately, the planners did not object and over time the two blended in reasonably well, helped by a few climbing plants on the walls.

An on-site project manager would have picked this up at an early stage but, as described above, we had failed to appoint one and our own visits combined with the surveyor's staged checks on the progress of the works were too infrequent to identify problems as they arose. One major mistake we made therefore was not to appoint a project manager at the beginning, although there are rather more sheep than human beings that far north in Scotland and project managers are few and far between.

Finally, we needed to point all of the external walls of the Mill building. In that part of the UK this involves snake harling; a process involving covering most of the stones with render, only leaving the prominent parts exposed. We had intended to be true to the original pointing but to add some cement to the lime and sand mix for strength, which we did later for the farm complex of buildings. However, in our absence, our builder substituted lime with a white-coloured cement mortar addition (Snowcrete®). This did not give the appearance of the lime–cement–sand pointing which we used successfully later in the farm complex projects and should have alerted us to the possibility of future problems.

The roof of the Mill was slate and in good condition apart from a few square metres of the funnel-shaped kiln end of the building. Rather than the fairly extensive reroofing specified by our architect, after close examination of the roof with our surveyor we reduced this to just the visibly necessary roofing works on the basis that the overall roof showed no evidence of water penetration and was likely to have been stable over time, so did not require extensive replacement. We needed to roof the new morning room extension and replace the existing galvanised metal flashing on the whole roof of the original building with lead. There are only limited types of slate used in that part of Scotland, so matching slates for replacement was not a problem. This was all straightforward.

- First-fix joinery

Before realising the potential problems in store, we had made the mistake of awarding the joinery works to the builder who had carried out the masonry works after, perhaps naïvely, attributing the two earlier problems to not adequately instructing the builder. However, we later found that he was not competent to do the joinery and could not complete the works. We then, on recommendation, found someone who not only corrected the problems left by his predecessor and took over the vast majority of the joinery but also project-managed for us from that stage in the works onwards.

1. Floor joists and other supporting timbers

All of these were intact in the Mill and we did not need any new supporting beams other than to fix joists in the former kiln drying/malting area in the smaller upper space in order to support the floor.

2. Floors

We used reclaimed engineered beech (Junckers®) throughout most of the main building, apart from whitewood in a couple of bedrooms we then carpeted. We managed to find around 150m² of reclaimed Junckers® engineered beech floorboards, in extremely good condition, which provided enough for the building apart from the parts which we floored with Caithness flagstones and the two rooms floored with whitewood and then carpeted. Junckers®, or similar, is readily available and is very high quality. It is light in colour, attractive and robust. In today's costs this came in at well under £20/m² and also avoided sealing/varnishing costs as it comes prefinished, a significant saving. None of this flooring work was complicated.

We used Caithness flagstones for the morning room, the small stair dogleg and for the two small areas on either side of the Mill workings. Bought from a specialist supplier these would have cost several thousand pounds for the area concerned, although courtesy of a local farmer (well, relatively local given distances in the north highlands) we were able to find the amount we needed for a couple of hundred pounds. These flagstones simply had to be cut to size and we laid them onto the concrete screed over the damp proof membrane. Hopefully none of the heritage of beautiful Caithness fields had been destroyed in order to obtain these; the farmer in question assured us that they had been stones lying around on the farm, which had been replaced by fencing decades previously.

The joinery works to convert the upper area are shown in **Photos 26–29**.

Photo 26 Mill. Upper floor to the smaller area with the heavy iron doors removed showing the entrance to the drying room.

Photo 27 Mill. Completed upper corridor with beech flooring into the new bedroom created in the upper kiln drying/malting room.

Photo 28 Mill. Finished bedroom in upper part of kiln. New windows, rooflights, cupola and light Junckers® flooring all add to the lightness.

Photo 29 Mill. The star-shaped beam structure supporting the kiln ceiling, stained by years of kiln smoke.

3. Windows and external doors

The windows and external doors were made by a local fabricator to match the originals, remnants of which were still in place, and fitted by a local joiner.

4. Staircases

Rather than a steel staircase, we had this made from Paraná pine and used tensioned galvanised steel wire for the staircase, tightened by screw turnbuckles at either end, rather than stainless steel as although the architect's proposal may have given a more striking result we felt that the huge additional cost was not justified. The original staircase had been wooden, and the galvanised wire was also perfectly in keeping with an old mill building. This made the staircase relatively transparent and at a fraction of the cost of either conventional wooden spindles or stainless steel fittings. The final cost of the staircase was therefore in the region of a few hundred pounds rather than thousands.

5. Dry-lining

We dry-lined the external walls of the Mill rather than rendering in order to add insulation, particularly as the north-facing lower main area had three large windows (and a glazed door) and the room would potentially be rather cold.

6. Insulation

As we were creating an essentially modern appearance, internally we plasterboarded over the joists on the lower floor and the ceiling beams on the upper floor, apart from the upper Mill workings area and the bedroom formed from the upper part of the kiln where the ceilings were vaulted to the apex of the roof. We used conventional fibreglass insulation for the flat ceilings and between the lower floor ceiling and the upper floorboards, as we were not limited by space, but used polyurethane insulation in the vaulted roof areas in order to achieve the necessary insulation with a far thinner layer (50mm) to keep the roofing structures visible.

As we had fitted plasterboard up to the roof apex in the new extension, we also insulated this section of ceiling with polyurethane boards to provide low thickness, high-grade insulation. (We left the new crossbeams of the A-frames exposed in the new extension and finished these with a dark, almost black, ebony stain which contrasted well with the white colour of the walls and ceiling).

7. Internal walls

These were either already present (the dividing walls between the large and small areas on each floor) or, in the case of the bedroom and bathroom areas, were formed with new stud partition walls.

8. Cupola

As a final first-fix solution in the Mill we also wanted to recreate the cupola, finishing the top of the kiln end of the building to return the appearance of the original working Mill. This process began with letters to some 30 Scottish distilleries with some success; the manager of one was kind enough to send me the plans for a cupola they had recently commissioned to replace their damaged original (a few indignant replies were also received, but these go with the territory – not everyone likes eccentric requests!). We used this as the basis of a plan for the local joinery fabricator for the Mill but his quote was prohibitive and we ultimately used our fabricator in County Durham, who produced exactly what we wanted with a copper canopy at a fraction of the cost of having this made locally. We then transported this to the Mill and installed it using a local crane. The cupola is not described in detail as this is more of architectural interest, but the structure matched those seen in distilleries around Scotland except that, rather than being open, the sides were made in the same louvre-style wooden construction in front of double-glazed panels; apart from the external appearance the cupola added light to the vaulted roof space. This whole process was remarkably inexpensive.

- Plumbing, heating, electrics, renewables and other secondary heat sources

Water and sewage plumbing were relatively straightforward parts of the project and, as described earlier, simply needed a septic tank with a soakaway and the necessary services pipework to be installed as part of the initial groundworks. The basic plumbing and electrics raised no problem because of the dry-lining of the external walls, flat ceilings and the stud partitions used in the upper floor in which we could run the necessary plumbing pipework and electrical cables.

In terms of heating, there were few options for oil boiler types at the time of the Mill project and although combination-type boilers were available they were still in their relative infancy and struggled to produce a sufficiently fast flow of hot water, so we opted for a conventional gravity-fed system. Likewise, condensing boilers were not in widespread use and we, perhaps unwisely, did not consider these at the time.

The electrics in the Mill were standard and raised no problems as we also laid these before fitting the flooring or installing the dry-lining and stud partition walls.

We did not attempt to use any sources of renewable energy in the Mill, because the concept of renewable energy did not occur to us (not being such a major issue at the time) and there would also have been few domestic systems available. However, considering the Mill from a modern-day perspective, the large open lower area would have suited a large (such as 7.5kW output) air-source heat pump which, as in later projects, would have greatly reduced our oil consumption and carbon footprint at low cost.

For secondary heat sources in the Mill we installed a multi-fuel stove with a back boiler plumbed through a simple gravity-fed system into a single large radiator, which together provided around 10kW of heat to the building. In retrospect, we missed the opportunity to make more use of the back boiler to heat additional radiators, which could have provided considerably more carbon-neutral heat given its potential 9kW output from the back boiler, in addition to the direct room heating from the stove. To repeat this project today, we would add a large, 7–8kW output air-source heat pump to a multi-fuel stove and back boiler to provide a potential output of around 22kW of heat. This would be sufficient to heat most of the building though an oil-fired central heating system would still have been needed for additional heat.

- Finishing joinery, other finishing works, decoration and finishing groundworks

1. Skirting boards, architraves and other finishing detail

The finishing woodwork parts were made from reclaimed wooden benches which we had bought in Glasgow and transported, together with a reclaimed cast iron spiral staircase, from Glasgow to the Mill via Durham, a fair number of carbon miles! The benches were made of a light-coloured hardwood, probably beech, and were machined in order to obtain the necessary parts.

2. Doors and door furniture

We made ledge and brace doors on site using wood cut from the same reclaimed benches. We then fitted the doors with simple new tee hinges and latches. Expensive replica old door latches and hinges could have clashed with the modern appearance of the finished building but we perhaps skimped rather too much in buying cheap standard modern door furniture and could have improved the overall look considerably by going a little more upmarket at limited extra cost.

3. Kitchens, bathrooms and other fittings

We found inexpensive and perfectly attractive, high-quality solid, light oak wood kitchen units with beech worktops together with a conventional modern white bathroom suite through online suppliers rather than using expensive specialist companies. These blended in well with.

4. Finishing groundworks

We needed very limited finishing groundworks at the Mill. We created a path from the parking area at the front down to the lower level and around the building and also added a largish semicircular area at the lower entrance to the house facing the river, all of which we covered with riverbed gravel and edged with traditional lengths of larch. In keeping with its location in fields, we simply laid the garden area to grass. We buried the relatively small amount of rubble created during the build below the path area.

An additional question was what to do with the existing areas of hard standing at the front of the upper entrance to the new morning room extension. This was flat and flush with the adjacent grass, about 6m in length and the width of the extension, but was cracked with a few small areas where the concrete had come away from the surface. The alternatives were to:

1. cover it with a fresh, shallow layer of concrete, although this would have been prone to erosion,
2. excavate down and add a fresh layer of poured concrete, which would have been extremely expensive and had a very new appearance,
3. add gravel or road planings – these would have gradually migrated onto the adjacent grass, or
4. leave it as it was.

We felt all of the options to change it were impractical and, ultimately, as the concrete was not in particularly bad condition, we decided to leave it as it was. Although rather ragged, the old and slightly cracked appearance probably fitted best with the building. Images of the building are shown in **Photos** **30–39**.

Photo 30 Mill. Finished open-plan area looking towards the fuel hopper.

Photo 31 Mill. Completed open-plan area looking towards the mill workings.

Case studies – the builds in practice

Photo 32 Mill. View to mill workings with open plan area and staircase detail.

Photo 33 Mill. Kitchen with cut kiln and supporting pillars for upper floor. The central island is inside the former kiln.

Photo 34 Mill. The ebony-stained timbers contrast the Caithness flagstone flooring and white interior.

Photo 35 Mill. As a break from emulsion, a few simple DIY stencils worked well.

Photo 36 Mill. Front view. The copper cupola and has aged well over the years.

Photo 37 Mill. Early appearance. Back view from the back.

Photo 38 Mill. Final appearance. Back view

Photo 39 Mill. A find on our quest for materials, by Lairg, Sutherland. That far north, 'by Lairg' covers a 30-mile radius!

The farm complex: common works to the Farmhouse, Barn and Byre

The farm complex consisted of five separate projects, all done at different times. We carried out the first three of these consecutively over a two-year period, although they were very different projects and each raised different issues. The design, planning, build specifications, process of finding builders and some of the masonry works were all common to the three initial projects (Farmhouse, Barn and Byre) and these stages are described below as a group rather than individually. Some of the principles we followed in these stages (notably the use of individual tradespeople) repeat throughout many of the projects.

Design and planning

There was little to be done in terms of design for these three projects as the Farmhouse had predefined rooms, we intended to leave the Barn and gin gan as open-plan spaces, and the conversion of the Byre appeared to be relatively straightforward and similar to the earlier modern Garage conversion though just on one floor. We had no intention of creating a dramatic 'statement piece' out of any of the properties, but to remain as close as possible to their original appearance. We felt that the existing window and wall openings provided sufficient light throughout the farm complex, with the exception of the barn where we needed to add two rooflights. We did need some specific permissions to make other minor changes to the fabric of the overall building complex. In particular, we needed to reduce the number of access doors (eight in the whole complex) and to rebuild two unstable chimney stacks in the Farmhouse. The other major changes were internal and were therefore more an issue of Listed building consent. These included creating a doorway from the existing Farmhouse through to the smaller area of the Barn, creating a doorway on the upper floor of the same area and rearranging the stud partitions forming the existing bathroom spaces in the Farmhouse cottage. We also created a second mezzanine mirroring the one on the upper floor of the smaller area of the Barn in the larger barn area and laid a damp proof membrane throughout the lower floor of the buildings. These proposals raised no problems in principle with the planning department or conservation planning officer. The only stipulation given to us was that we could not divide the original buildings into separate living units, something that we did not intend to do at the time.

In terms of drawing up the relevant plans for the buildings in the farm complex, with the experience of the Mill under our belts we changed tactics with the services of an architect. We did instruct one, but this time based on a fixed fee basis to provide drawings and elevations from our own designs for the buildings but not to go beyond this stage other than to provide occasional advice. We submitted these plans and obtained planning permission and Listed building consent without any difficulty. Building control specifications were relatively straightforward as most of the general building control requirements were included in the building control submission, so this whole process was simple. We did not go out to tender for the works as we were fortunate enough to have builders we trusted from previous experience.

Build specifications

Rather than drawing up detailed specifications in advance of the build, which we would have needed for a tendering process, we worked backwards from the building control plans and agreed the specifications with our builders as we went along.

Finding builders

We were fortunate in that we already knew our builders (The Boys) for these three projects. We did not, therefore, have to look far. We already had full trust in their knowledge and approach to building works and so we did not go through any tendering process. The Boys were also able to recommend dedicated tradespeople (roofer, electrician, plumber etc.) with whom they had worked and we similarly did not look any further for most of these tradespeople.

The builds themselves

- Initial groundworks and services

The groundworks were relatively simple for the farm complex projects and involved digging a services' trench to the farm for water and electricity, which we left open for a short time during the build for our archaeologist to document the earlier foundations we found. We also needed a short trench to connect the Byre sewage to the main system in the courtyard, all done manually by The Boys' labourer.

- Masonry works, roofing and guttering

1. Masonry works

The common work for the whole complex was the external pointing, a huge potential cost, and one which can easily be overlooked. The major choice for this, in terms of both materials and the ease of pointing (and therefore building cost), was the type of pointing materials we used. A limited amount of the old lime pointing was still present, but the vast majority of the building, with a perimeter of some 250m and average height of around 4m, needed complete repointing.

Our walls were constructed of separate inner and outer walls built using lime mortar and filled with a small amount of loose stone rubble, which we felt would behave more like a cavity wall than a solid one. We did not, therefore, seek to use breathable layers throughout the thickness of the wall but pointed with a '1:2:6' mix (one part cement, two parts lime and six parts sharp [coarse] sand) made on site. We felt this offered the optimal compromise of strength, appearance and breathability as opposed to water-impermeable properties, for our wall structures.

We used bags of hydraulic lime in our pointing mix rather than the slower-setting non-hydraulic lime, commonly sold in the form of a lime putty, which many experts will prefer; the latter is far slower to set and more time-consuming to apply (see this article: Lime, Hydraulic or Non-hydraulic?[10]). Likewise we did not use lime-render to finish our inner walls, but used either a cement-based render or dry-lining. It is essential however to obtain sound advice when deciding on the optimal materials to use to treat internal and external walls and solid lime-mortar walls demand a quite different approach, avoiding cement to allow the lime in the wall to act as a 'wick' to enable water vapour to pass through the wall rather than trying to reduce water penetration in the first place.

Many old lime mortars may also contain additions, in our case flecks of coal and small fragments of gravel, probably from the local sand used in the mortar at the time. We added some crushed coal fragments and coarse grit to our pointing mix in order to optimise the match with the original lime mortar, which produced an excellent appearance. With the added strength of the small amount of cement, this is still in perfect condition 20 years later (lime mortar is not the most robust of materials and may require more frequent repointing – a significant maintenance cost). It has also mellowed further in colour over time (see **Photos 40–42**). Whilst we have had small amounts of sand and grit falling internally, presumably as the walls have dried out, this has occurred on the external and the internal facing walls, so was not a result of inadequate wall 'breathing'. We resolved this easily with a gentle brush down, a run over with Henry the Hoover (see anecdotes) and a thin spray of a proprietary stabiliser,

Case studies – the builds in practice

not to waterproof the wall but just to bind some of the sand and salt which might fall in the future.

Internally, apart from the colder north- and east-facing walls of the farmhouse which we dry-lined, we used a conventional sand and cement render. These reduced the materials and labour costs to a fraction of what this would have been using lime mortar.

The concept of walls 'breathing' is complicated and many documents are available describing the circulation of water vapour in walls depending on the type of internal and external pointing or wall coverings. In order to remain true to the concept of a wall breathing the materials used will, at least theoretically, need to be breathable right through the wall. This needs the internal walls to be either pointed or finished with a lime mortar and breathable paints. The same applies to dry-lining, which would need to use a breathable material rather than conventional polythene. The Boys' estimate that using lime putty-based products in their current work on a World Heritage site has greatly increased the cost of pointing and rendering.

In practice, the choices we made have not led to any internal damp, condensation or detectable

Photo 40 Farm. The original Farmhouse walls needed extensive repointing: the existing build structure dictates how this can be done.

Photo 41 Farm. The same wall 20 years later: no internal stonework deterioration or dampness.

Photo 42 Farm. The whiteness of the aged lime mortar (the small section below right) stands out more against the stonework than the wet mix containing just a small amount of cement.

deterioration of the external walls. These decisions appear, therefore, to have been successful and reduced our costs across the three projects by some tens of thousands of pounds, in addition to reducing maintenance costs compared to using a proprietary lime alternative.

2. Roofing and guttering

Reroofing the existing Marley-tiled Farmhouse roof with clay pantiles would have cost well over £20,000 in today's money and would also have clashed badly with the older tiles on the remainder of the property. Realistically, the only solution to avoid this would have been to reroof all of the connecting buildings making up the farm complex, excluding perhaps the outbuildings. This would have involved around 600m^2 of roof which, if the average cost on websites of £100/m^2 is to be believed, would have represented a present-day cost of around £60,000 for the whole complex of buildings. New pantiles would also have stood out far more than the combination of the existing concrete tiles and the old pantiles and would have given the whole property a 'modernised' appearance rather than keeping the original appearance of the buildings. On the advice of The Boys, and with the agreement of the conservation planning officer, we therefore settled on a repair solution: finding the few additional old pantiles and Marley tiles needed to repair the few small damaged areas of roof and replacing the Byre roof, which was in a poor state. This left the new appearance as close as possible to the original.

- First-fix joinery

As the projects were all different, both in terms of the initial structure and the type of build (restoration, conversion and old versus new final appearance), we had few specific first-fix works common to all – other than the use of a local joinery fabricator to construct the windows and other joinery materials.

- Plumbing, heating, electrics

Again, we had few common works to the plumbing other than the types of heating (oil) and our use of polypropylene (Hep2O®-type) piping throughout these three initial main farm complex projects. Likewise, we used the same type of heating systems: conventional boilers with gravity-fed systems. Also and again like the Mill, in retrospect we could and probably should have used condensing boilers. We decided against combination boilers. Whilst these are more efficient in terms of hot water, they only offer limited savings compared to the cost of space heating and in our experience in the early gin gan project (not described), they struggled at the time to produce enough hot water fast enough for our uses.

- Finishing joinery, other finishing works, decoration and finishing groundworks

Finishing joinery varied between the three projects as we had different intended final appearances. Kitchens and bathrooms are not described in any detail as these are personal preferences, although all of ours had the common feature of being simple in order to remain in harmony with the finish of the different buildings. We obtained these from online retailers at a fraction of the cost of those from specialist suppliers.

The finishing groundworks throughout the farm complex were relatively simple, although involved slightly more than at the Mill. The stables outbuilding was approximately 1.5m higher than the building level and we had to carry out some groundworks in order to create an even slope down to the path around the buildings. Because of the slope, lengths of larch would not have adequately held back the earth before it had been stabilised with grass, so we used leftover stones from the build to line the bottom of the grass, also with a modest cost saving, but a saving nonetheless. Once bedded down these looked attractive and cost nothing apart from a morning's work by a labourer to lay them.

We managed to lose a fair amount of rubble as a base for the parking area, driveway and paths around the buildings and covered the large parking area and driveway with road planings. These compressed down easily after being driven over a few times by the JCB which spread them and gave a robust finish which, although not necessarily the most attractive, was visually unobtrusive and consistent with what might be used on a farm. Road planings are a cheap and robust option (a 20-tonne load costs around £300 in today's money). If their appearance upsets anyone's sensibilities they can easily be covered with a layer of gravel – four or five tonnes are sufficient to cover an area to accommodate half a dozen cars. We did this later on. We then created simple paths from the parking area around the buildings and covered these with gravel. Apart from costing several times as much, specialist gravels would have looked out of place around the buildings making up the farm complex.

We spread around 300 tonnes of assorted plasterwork and rubble generated from the works over the grounds, saving the high cost of disposal. Disposal costs are easy to overlook and, although we did not research these, one website suggests that they may amount to some £20/tonne (see Homebuilding & Renovating article[10]). Not everyone will have a suitable place to fill with rubble, but assuming there is a sufficient ground area around the property it is many times cheaper to spread it over the grounds and cover with topsoil than to have it taken away.

The hard standing area at the farm complex is the courtyard formed by the Farmhouse, Byre, small set of outbuildings and the Barn, and slopes down to an entrance gate. It had a total surface area of around 100m² and was an interesting jigsaw of existing concrete surface, areas of erosion, holes, other cracks, and filled areas from the couple of service trenches we had dug. The slope made any form of gravel impossible and excavating down sufficiently to allow a decent thickness of concrete and/or asphalt to backfill would have cost a fortune in labour and materials. It would also have meant disposing of at least another 100 tonnes of rubble which we had no further space to use. Regardless of the huge costs involved, either process would have produced a new finish which would not have sat well with the original appearance of the Farmhouse, Barn and Byre complex of buildings. Any form of modern thin 'spray on' surface coating would also, potentially, have looked completely out of place, would not have stayed the course of time given the cracked concrete base and would have been unviable and expensive. We decided that the patchwork finish to the courtyard, although not ideal, was probably the least poor option and have simply allowed wildflowers to grow through the cracks in the cement, giving an attractive, if rather eclectic, appearance, not unlike the areas around the rest of the farm complex. This may be a useful way of avoiding unnecessary cost and leaving an unchanged appearance in a conversion or restoration project.

We have a 2 hectare field (just under 5 acres) next to the farm grounds, part of which contained a slight dip around 20m long and 10m across. The base below the topsoil was clay and was therefore suitable for a pond without any form of lining. With one day of a JCB and a small grant kindly given by a local bird club, we were able to create a pond that has acted ever since as a magnet for swallows, swifts, flycatchers, geese, ducks, coots, kestrels and a heron – although for the heron it is a triumph of hope over experience as there are no fish in the pond: we have kept it fishless to support invertebrates, and therefore birds (see Photos 138–139 on Pages 191 and 192). For us, the arrival of our five pairs of returning Greylag geese heralds the start of spring. If suitable space and subsolum is available, this is a cheap but highly-effective and attractive piece of finishing groundwork which also makes a decent contribution to nature conservation.

- Renewables and secondary heat sources

In terms of renewables, we considered the main options for these throughout the Farmhouse, Barn and Byre some five years after the original build, when they were in more general use.

1. Wind energy

The electrician who provided our energy performance certificate (EPC) advised that a wind turbine (estimated cost £15,000–£20,000) would provide annual savings of around £300–£400 in the three farm buildings, or a return of around 2–3%, with a 30-year payback excluding depreciation. When we asked a company to provide us with an estimate a few years ago we were told that as the Farm is located at the bottom of a hill we would experience turbulence despite having no houses in the vicinity. In addition, and again this is not included in average wind speed estimates, we tend to have either no wind or a gale that would exceed the maximum tolerated speed of the turbine, and the return would therefore be somewhat less than the estimate of 2–3% based on average wind speed.

2. Solar energy

We have a south-facing pantile roof on the barn part of the farm, sufficient to house around 10 solar panels. An initial approach to planning was not met with much enthusiasm, so consent would have been an issue; although with a little persuasion we might have been able to resolve this by positioning the panels elsewhere. The return on the £5,000–£7,000 investment would have been around 3–4% excluding depreciation, assuming that we managed to optimise our use of the electricity generated, as the feed-in tariffs to the National Grid were so low.

3. Ground-source heat pumps

A ground-source heat pump was not really an option for us because these only came into more widespread use several years after we had completed the projects on the buildings making up the farm complex and the prospect of ripping up floors to install the pipes or lining the walls with large low-flow radiators was not attractive. In addition, the nature of the building and its rubble wall construction is not well-suited to the low-level background heating produced by most water-based heat pump solutions. As discussed later in Chapter 5, the air-to-air heat pumps proved to be an effective option in our case, and probably for many others.

4. Biomass boilers

We looked briefly at these but after a quotation of around £30,000 dismissed them on the basis of both cost and return on investment but also because we already had extensive secondary sources of heat which were either carbon neutral (fires) or low carbon (electricity for the heat pumps).

5. Air-source heat pumps

After dismissing air-to-water heat pumps designed purely to heat water, as this accounts for only a small fraction of our heating costs, we briefly considered air-to-water central heating systems but rejected these essentially for similar reasons as the ground-source heat pumps.

We concluded that the air-to-air heat pumps offered compelling advantages over all of the other renewables (both for heating and electricity generation), were both highly cost-effective and would greatly reduce our carbon footprint compared to running a central heating system alone.

An overview of how we used both air-source heat pumps and multi-fuel stoves is given in the next chapter and shows the benefits of what we decided to use.

The Farmhouse

This section describes the restoration of a completely derelict but structurally sound period farmhouse. Design and planning, build specifications and finding builders were all common to the three initial farm projects and are described above. The descriptions below relate, therefore, to specific aspects of the Farmhouse project.

The build itself

- Masonry work, roofing and guttering

The masonry works in the Farmhouse were not particularly complicated, although some decisions had a significant effect on both cost and

Case studies – the builds in practice

the final result. Structurally, we needed to remove around 1.5m height of two unstable chimney stacks down to stable brickwork and rebuild them. The conservation planning officer, an extremely useful source of advice, was helpful in advising on the best replacement bricks to choose from the multitude available

We needed to close a couple of the six separate entrances to the Farmhouse: these included four to the Farmhouse itself and two to the smaller Barn space which was to become the Farmhouse kitchen. Two appeared to be unnecessary and we closed these, leaving the existing door as an external façade on one side of the Farmhouse and painted wooden panelling on the courtyard side.

We also needed to knock through the walls to create two new internal doors, one from the Farmhouse into the lower smaller area of the original barn and the second from the mezzanine above this area of the original barn to access what would be the mezzanine in the new Barn build.

Finally, or as at least we thought at the beginning, as the walls coverings in the Farmhouse (lath and horsehair plaster) were all weak or falling off, these needed to be stripped down fully and either rendered (most of the walls) or dry-lined (notably the coldest north- and east-facing walls).

It is potentially easy to underestimate quite how much work this can involve and how much rubble it will generate.

Two approaches are available for damp proofing floors:

1. excavating down, laying a membrane and covering this with a concrete screed, or
2. laying the membrane on the existing floor and then covering it with cement.

We used the first approach in the Farmhouse, as the latter would have involved raising all of the ground floor lintels at a huge expense and the need for Listed building consent. In practice, and with the help of some exploratory digging, this was not a difficult process as the existing concrete floor was thin and easily removed, leaving loose hard core and earth beneath. We did need to raise one lintel.

In terms of roofing and guttering, the Farmhouse roof was sound apart from the ridge tiles, which we needed to replace; we found replacements in a local demolition yard together with a few tiles to patch the few damaged roof areas. Whilst the original Farmhouse roof had been replaced with brick red-coloured, concrete Marley tiles, these had aged well over time and the overall appearance of this roof was surprisingly similar in colour to the pantiles elsewhere on the farm roofs. The old aerial photograph from 1964 (Photo 20, Page 62) shows the Farmhouse with a pantile roof and so the Marley tiles were presumably added in the period between 1964 and the time the Farmhouse stopped being occupied. The conservation planning officer's personal preference was to reroof the Farmhouse building, but he was not insistent and later agreed that our decision had been correct in terms of preserving the original appearance.

Guttering (a mix of wood and cast iron on different walls) was straightforward to replace. We avoided the sizeable cost of the 60m of wooden guttering needed for the Farmhouse (and Barn) by a phone call to a local demolition yard, which was very happy to offload all we needed for a few pounds.

As the work progresses further restoration challenges came to light. **Photos 43–46** shows the most demanding of these. The section of wall shown was the most interesting of our architectural finds and needed the most sensitive approach. The most obvious would have been to appoint a company specialising in the restoration of old architecture. However, we opted for a simpler approach of taking advice from the English Heritage officer about final appearance and on possible repair

Photo 43 Farmhouse. More of architectural interest: the stone quoins uncovered in the gable end of the former Farmhouse.

Photo 44 Farmhouse. Initial appearance of the medieval window found above the existing ceiling.

Photo 45 Farmhouse. The medieval window after restoration and masonry work to strengthen the gap in the window.

Case studies – the builds in practice

Photo 46 Farmhouse. Some of the other back of the window remained (plastered over): careful sandblasting avoided damaging the stonework.

options from our builder we repaired the wall. This was fairly straightforward and involved inserting stone into the window space, creating a central pillar and strengthening the slight lean in the upper wall by knitting some stonework together.

The second set of illustrations (**Photos 47–51**) photographs shows the lower wall and progression of work from stripping off old plaster revealing different styles of stonework in the wall. Removing the stones which closed the small opening at one side revealed the large cut stones held together with wattle and daub. Gentle sandblasting and brushing down of the lower surface revealed a wall with its long history, showing the progression of the life of the building. As the cleaning process proceeded, the heritage of the old wall, originally covered by plaster, became increasingly obvious. Again, working with some specialist advice and careful thought rather than instructing a separate specialist company considerable cost and time was avoided

Photo 47 Farmhouse. Part of the lower gable end wall showing evidence of a closed up section.

Photo 48 Farmhouse. Careful removal of the wall covering revealed the opening. Easily missed.

99

Photo 50 Farmhouse. The whole length of the central lower gable end of the Georgian section.

Photo 49 Farmhouse. Repair to the stonework (same 1:2:6 cement, lime and sharp sand mix) and exposure of the woodwork in the opening.

Photo 51 Farmhouse. 750 years of history in a wall. The lower gable end shows the progression from medieval times to the post-Victorian era.

restoring the wall to show the different ages of the sections. We were reassured by the reaction of the English Heritage office to the final result that the decisions we had taken were appropriate. In addition to the main finds, several small sections of wall containing closed up old openings came to light (**Photos 52 and 53**). These could easily have been missed as the stonework closing the openings was similar to the existing, and perhaps illustrate the need for extremely careful inspection.

The full history of this part of the house remains unclear, and all we could do was save what was there in the hope that we might gain some enlightenment later.

Photos 54–69 show progression of the works to the finished result throughout the Farmhouse building.

- First-fix joinery

In the Farmhouse project, we often had to remain one step ahead of our builders in terms of obtaining materials, particularly reclaimed ones, to avoid holding up the building works; unlike a modern build where new materials are simply bought from local building supplies merchants. A fair amount of our first-fix joinery materials came from the weekly local free advertising paper or from reclamation yards.

Case studies – the builds in practice

Photo 52 Farmhouse. The uncovered walls revealed another inset which had been closed with stonework and plaster.

Photo 53 Farmhouse. One of many small features found by careful inspection of exposed walls.

Photo 54 Farmhouse. Middle upper room giving a better view of the original older roofing timbers.

Photo 55 Farmhouse. Middle upper room wall with covering removed shows the lintel from the original opening and stonework.

Photo 56 Farmhouse. Middle upper room. Finished roof timbers showing pegged joints and slightly convex pegged arms of the A-frame.

Photo 57 Farmhouse. Middle upper room. Finished gable end and exposed timbers with the vaulted ceiling.

Photo 58 Farmhouse. Gable end upper room with fireplace stolen. Initial appearance.

Photo 59 Fireplace. Gable end upper room after stripping out before dry-lining, revealing the original chimney breast.

Old Buildings: Conversion and Restoration

Photo 61 Farmhouse. Middle upper room: the more recent stonework was unremarkable and rendered, leaving the chimney breast exposed.

Photo 60 Farmhouse. Same room with finished chimney breast.

Photo 62 Farmhouse. The fireplace area in a lower, bedroom in the Georgian end. Initial appearance.

Photo 64 Farmhouse. Above the brickwork, very old notched timbers suggest rebuilding dating back centuries. Left exposed.

Photo 63 Farmhouse. The same fireplace in a lower room sandblasted and repointed.

Photo 65 Farmhouse. Central room with exposed upper floorboards: the offending modern fireplace before removal.

Case studies – the builds in practice

Photo 66 Farmhouse. Excavated fireplace in central room. 'New' stove with back boiler and blackened stonework from a few centuries ago.

Photo 67 Farmhouse. Winter warmth. The stove and its two radiators are sufficient to heat the whole core of the house without burning oil.

Photo 68 Farmhouse. A vintage Rayburn drank oil but at least kept bottoms warm.

Photo 69 Farmhouse. Replacing a thirsty wick burner oil Rayburn with a woodburner provided more heat at a fraction of the cost.

1. Floor joists and other supporting timbers

Relocating the staircase from the smaller to the larger part of the Barn in order to create a kitchen in the smaller area left a hole in the floor, two joists wide.

We needed to find two replacements, of the same thickness and depth, rather than trying to stain modern whitewood or plasterboard over the attractive underside of the above flooring and the floor joists.

This took a couple of trips to local reclamation yards but it is now impossible to tell the original from the replacements.

2. Floors

We used all of the possible options in the Farmhouse. We laid modern whitewood in two lower rooms, which we then carpeted. We also carpeted one of the upper rooms in which the original boards were structurally sound, but unsightly after we had filled the woodworm-damaged surfaces. We left the original floorboards exposed in the other two upper rooms and laid reclaimed pitch pine in the central lower room. The floorboards in the central upper room, which were around 25cm wide, were slightly damaged from woodworm infestation: after we had gently sandblasted them they appeared to be repairable with only limited filling (using a sawdust and glue mix), leaving an attractive finish with a few small patches filled and stained to the colour of the boards themselves. In the kitchen area, we laid 250mm wide Douglas Fir pine boards cut from floor joists (see Barn section). The boards had imperfections from old cuts and nail holes, although alongside the exposed external walls these imperfections, combined with the age of the wood, gave it an attractive patina and a good final appearance which was in keeping with the room. In our view, this gave a far better appearance than any new hardwood or high-quality softwood, but at a fraction of the cost.

There were one or two other gaps in the exposed ceiling where the upper floorboards had been damaged. Some showed evidence of previous repairs. We simply covered these in the same way as identifiable earlier repairs by fixing narrow stained battens onto the lower ceiling (the previous repairs had actually used short thick leather straps). Although perhaps not the most refined of solutions, this type of repair is both extremely simple and true to the techniques used in the history of the building. It also made our conservation planning officer happy!

All of the undersides of the upper room flooring were attractive and different in appearance: old, probably Georgian era, joists and boards in the middle room; far older joists on the other upper room of the main Farmhouse; and joists and floorboards dating from the Victorian era in the cottage end. Although leaving the ceiling boards exposed in the lower rooms of the Farmhouse meant that we could not install heat and sound insulation in the ceilings, we felt that the trade off against the attractive finish justified this. Whilst we made these decisions based on final appearance rather than cost, we did also make savings from keeping and repairing rather than replacing the flooring in the upper rooms, particularly given the cost of buying wide floorboards and plasterboarding over the other lower ceilings.

3. Windows and doors

The windows in the Farmhouse were sash type; five with eight panes in each sash in the front of the Georgian end and either one- or two-pane sashes in the Victorian cottage end. Three small windows on the back of the Farmhouse, in the long extension and probably added in Victorian times, were also simple single- or double-pane construction (see **Photos 70 and 71**).

To pay homage to our joinery fabricator, he replicated these accurately at a cost of around £400 in today's money for the more complicated 16-pane windows. A brief internet

Case studies – the builds in practice

Photo 70 Farmhouse. The most complicated replacement Georgian sashes produced by our fabricator: indistinguishable from the originals.

Photo 71 Farmhouse. The Victorian cottage end. Retaining two different window styles was essential to show the history of the building.

search will reveal that most specialist companies do not even offer an indicative price for these, which is always an ominous sign.

4. Staircases

Excluding the staircase in the smaller Barn area which was to become part of the Farmhouse later, the Farmhouse itself had two staircases.

One staircase dated from the Georgian era and was slightly worse for wear as a result of over-enthusiastic woodworm and some of its treads were cracked. It was nevertheless attractive and would have been difficult, visually incongruous and hugely expensive to replicate and replace. Overall, the staircase was sound apart from some superficial damage and so we chose to leave it alone apart from gentle sandblasting, brushing and resealing in order to preserve its original appearance in the building. We did not cost the replacement option because we wanted to keep the original, although given the shape and complexity of the staircase the savings probably amounted to a few thousand pounds. Many people will consider sandblasting a technique used primarily to clean stone and metal. However, applied carefully and gently on site, we found it to be immensely effective on wood. Apart from its standard application on stone and metalwork, we used a sandblaster to clean some architraves, the Georgian staircase and various supporting timbers and flooring, where it left a similar surface to machine sanding but was more effective on the irregular, uneven depressions. Although the lightly sandblasted and varnished flooring was not an ideal finish, we chose this over replacing the floorboards, firstly because the degree of damage did not really justify replacing 60m^2 of extremely expensive wide flooring, and secondly to keep the original appearance of both the upper and exposed lower surfaces of the boards forming the ceilings of the lower rooms.

The second staircase was in the

Victorian cottage end and dated from that era. This had been painted and although we could have stripped off the paint, we simply carpeted it as it was otherwise unremarkable.

5. Dry-lining

We rendered most of the walls in the Farmhouse with a sand-cement mortar mix and have not noticed any particular cold or damp as a result of our decision not to dry-line all of the walls. I have discussed the issues around walls 'breathing' and types of materials used earlier. Dry-lining was limited to the north-facing walls in the Farmhouse where the additional insulation was useful to protect against heat loss from the cold wall areas, although we simply rendered the east-facing wall, as the extension spanning most of this wall provided a heat buffer protecting the main part of the Farmhouse.

6. Insulation and ceilings

We were fortunate in being able to reduce our insulation costs enormously by finding around 600m² of reclaimed 25mm polyurethane foil-lined sheets (Kingspan®-type equivalent) for a couple of hundred pounds. Although a bit grubby and slightly ragged at the edges in places these were easy to trim to size and took over £6,000 off the total potential build costs to no detriment.

7. Internal walls

The internal walls throughout the Farmhouse were made of solid stonework apart from a rather contrived stud partition arrangement in the Victorian extension running the length of the building. We removed these partitions and rearranged the area to create a bathroom and lower hall in the Victorian cottage end. Some examples of repairs and replacements are shown in **Photos 72–81**.

- Plumbing, heating, electrics, renewables and other secondary heat sources

In the case of the Farmhouse, water drainage was already in place and both dirty and clean surface water ran into the Victorian pipework, which in turn led into the septic tank. We needed to install internal plumbing for the two additional bathrooms in the Farmhouse part of the build and the kitchen; this was relatively straightforward as we installed all of the plumbing infrastructure before laying the flooring in rooms or dry-lining any walls. In the Farmhouse, we ran the upper floor plumbing pipes in box skirting board structures running along the walls using the Douglas Fir boards. Otherwise, the plumbing infrastructure in the Farmhouse raised no problems.

In terms of heating, we had few options for boiler types at the time of

Case studies – the builds in practice

the farm complex projects and although combination type boilers were available they were still in their relative infancy and struggled to produce a sufficiently fast flow of hot water, so we opted for conventional gravity-fed systems. Likewise, condensing boilers were not in widespread use and so we did not consider these at the time. We needed two boilers in the overall farm complex because of the overall length of the building, being some 60m from one end to the other around the U-shape of the four component parts (Farmhouse, Barn, its gin gan and the Byre); one boiler for the Farmhouse and one for the Barn, its gin gan, and the Byre.

Photo 72 Farmhouse. Two joists added after removing a staircase: the reclaimed additions are indistinguishable from the originals.

Photo 73 Farmhouse. Underside of the middle Farmhouse upper room left as the exposed ceiling to the lower room.

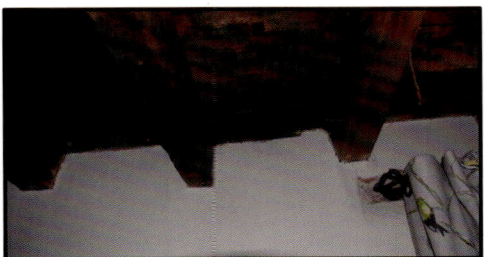

Photo 74 Farmhouse. Damaged sections of the underfloor: old repair on right (leather strap) and same style repair (stained batten) on left.

Photo 75 Farmhouse. The cleared upper surfaces: not perfect but the damage marks do not detract from their overall the appearance.

Photo 76 Farmhouse. The original Victorian bedroom fireplace and surround.

Photo 77 Farmhouse. Sandblasting combined with a great deal of hard work stripped this fireplace back to the original.

Old Buildings: Conversion and Restoration

Photo 78 Farmhouse. The initial appearance of the Georgian staircase, at first sight beyond redemption, worth looking at more closely.

Photo 79 Farmhouse. Same staircase after sandblasting. The wood indents made any alternative cleaning method pointless.

Photo 80 Farmhouse. The staircase finished and sealed; replacement was unnecessary and the repair provided a good final result.

Photo 81 This young lady came from a small island off Guadeloupe via foot, ferry, car, taxi, plane, second taxi and plane, then car to Durham; all with a toddler with incipient chickenpox and Gillian pregnant with twins!

Case studies – the builds in practice

One of our poorest decisions was with the plumbing infrastructure and, as a result, we have had repeated problems from rodent damage to both electrical cables and plumbing pipework. Rather than using the conventional approach of lengths of copper pipe soldered together, we opted for the plastic Hep2O®-type pipes, which are 'push fit', easier to join, and can be curved into position rather than needing to be soldered. Unfortunately, these pipes also seemed to act as a magnet for rodents and caused us no end of problems.

It is almost impossible to completely avoid rodents from getting into old rubble-filled walls and it might be wise to consider some armoured cabling or conduits in vulnerable areas for electrical cabling and to avoid plastic plumbing pipework. The old adage that we are never more than six feet away from a rat was never truer in this case!

After various leaks from rodent damage, including the consequent oil and water bath described in the anecdotes in Chapter 6, we finally resolved at least the majority of the problem by pouring a concrete screed and tiling over the main pipe run in the Farmhouse; something which, in retrospect, we should have done in the first place.

We did not install any renewables at the time of the original Farmhouse build but did add several to the various buildings in the whole complex some five years later. I have described these later on in the Barn, Byre and Farmhouse Extension projects.

We did decide to use two secondary heating sources:

1. A reclaimed Coalbrookdale Severn, a marvellous multi-fuel stove with a back boiler that has worked perfectly for over 20 years. This was one of the best investments we made and cost us around £100 today's money plus a bit of a lad's time picking it up. Historically, the Coalbrookdale stoves were top-of-the-range, attractive and easily adjusted. They probably still are, although the choice of stoves is far greater. The stove was plumbed into three radiators (in retrospect we could have plumbed it into more, as it has a rated back boiler output of over 10kW). The radiators and stove together heat the core of the Farmhouse in all but the coldest weather.

2. A solid fuel-burning Aga in the Farmhouse kitchen, which we plumbed into a single radiator.

The net result of these systems was that the two plumbing 'control hubs' looked a little bit like ship's engine rooms, but they have both worked very well, again for over 20 years. The Aga proved less successful, as we found that in order to keep the beast burning or to produce sufficient heat for cooking it needed regular feeding and we soon tired of the two-hourly trips to the coal shed. We subsequently replaced it with an old wick burner oil-fired Rayburn, but this almost doubled our oil consumption. Far more recently we replaced this with a DEFRA-approved 6kW Mazona Rocky fire, which has provided generous heating for the large vaulted ceiling kitchen at a capital cost of around £300 including a few accessory flue parts and much the same to install (given that we had an existing chimney). This fire comfortably paid for itself in its first year of use compared to the old thirsty Rayburn.

The electrics in the Farmhouse were uncomplicated and did not raise any issues other than a major problem we have experienced over the years: the rodents! It was an expensive oversight not to consider these four-legged beasties.

- Finishing joinery, other finishing works, decoration and finishing groundworks

The main issue with the finishing joinery in the Farmhouse was ensuring that we had the necessary wood available in advance in order to replace badly damaged door jambs, architraves and skirting boards and to make replacements for those doors which were missing or badly damaged. We were very successful in doing this using wood reclaimed from various sources, although as can be seen later in Photo 88, taking an 'eye-off-the-ball' can leave some finishing joinery, in this case woodwork above a door, clashing badly with the appearance of the neighbouring joinery. We acquired most of our reclaimed wood for pennies and the main cost was the labour required to make the doors (described below) or machine the reclaimed wood.

1. Skirting boards, architraves and other finishing detail

These were all machined from the reclaimed planks and posts obtained from our finds along the way or from the machined Douglas Fir floorboards.

2. Doors and door furniture

We made most of the doors needed in the Farmhouse on site from the pitch pine boards. These were again of ledge and brace construction consistent with the originals. They took little time to make and were a near-perfect match to the remaining existing doors at around 20% of the cost of similar commercially-available doors, which may not have matched the originals nearly as well.

Case studies – the builds in practice

We were fortunate in having door furniture on most of the existing doors and on the outhouses although much had rusted badly over the years. These were not the most delicate of parts but very much exemplified the door furniture used in farm buildings at the time. We recovered these and, after a bit of research, I carried a few boxes to a local sandblasting company (thank you Cyclex Ltd, Shildon, Co Durham) which kindly put them in a corner of the blasting chamber and stripped them back to the bare metal. We then coated these with a wax/turpentine mix to bring out the appearance of the metal and to prevent them from rusting.

Some examples of retaining original woodwork are shown in **Photos 82–88**.

3. Kitchens, bathrooms, decoration and other finishing details

We had a lucky find for the Farmhouse kitchen with some 20 small pitch pine doors and panels from a local wood stripping company, which were almost exactly the size of modern kitchen unit doors. We simply bought some cheap, robust secondhand kitchen unit carcasses, trimmed and fitted the doors and end panels and added brass handles. The overall cost of buying these was negligible and the joinery required was rather more which

Photo 82 Farmhouse. Lower Victorian room. Original plasterboard ceiling removed: joists and underside of upper floorboards exposed.

Photo 83 Farmhouse. The final appearance of the lower Victorian room, with exposed joists and timbers.

Photo 84 Farmhouse. The same range, cleaned and finished with Zeebrite®. A wonderful relic from Victorian days.

Old Buildings: Conversion and Restoration

Photo 85 Farmhouse. An original door. Slightly moth-eaten, but fine in an old Farmhouse.

Photo 86 Farmhouse. A new door with reclaimed sandblasted hinges and latch, very close to the original.

Photo 87 Farmhouse. An external door: there is absolutely no need to replace many original doors.

Photo 88 Farmhouse. Eye off the ball. The reclaimed wood above clashes dreadfully.

meant that the overall final cost was probably not very different from installing relatively inexpensive new units, but the older style fitted in well with the exposed walls and fireplace in the new kitchen area.

Apart from some eccentricities in acquiring secondhand bathroom fittings from a range of remote places around the county, all of the finishings were unremarkable and kept simple.

We also kept the decoration as simple as possible. The only difficulty we encountered was our experimentation with traditional casein-based paints. At the time these all used a powder paint base and natural earth pigments. We thought this would give a more traditional appearance and allow the walls to breathe – an incorrect belief on both counts as we had rendered the walls with an impermeable sand/cement mix and in practice it is possible to replicate the colours with modern emulsions. It is, I understand, now possible to buy these pre-prepared in tins, although at the time they relied on adding a precise amount of pigment to the powder base. We would be very cautious about using these again unless this was really necessary (preserving breathing in a lime mortar wall). At the time, and I suspect still, the casein-based paint did not store well and tended to go 'off' and because of the way these paints were mixed, it was also not practical to prepare small amounts of paint for touching up. Less 'eco-friendly' impermeable modern emulsions can reliably colour match for touching up purposes and store well: a tin bought over 20 years ago is still in use. However, the real problem with casein-based powder paint came when we decided to repaint. The powder-based paint makes repainting impossible without a coat or two of stabiliser, which made redecorating a room a complicated, messy and expensive process.

The finishing groundworks are described in the common works section.

Photo 89 Farmhouse. Reclaimed pitch pine panels used for the doors on reclaimed carcasses.

Photo 90 Farmhouse. The final appearance of the kitchen; rather busy but it works.

Photo 91 Farmhouse. Reusing the existing bathroom fittings posed no problem and looks fine.

Photo 92 Farmhouse. After sanding our behinds, this bath found its place back outside: a case of over-zealous reclamation!

Photo 93 Farmhouse. Finished appearance of the front of the whole Farmhouse.

Photo 94 Farmhouse. Home and Happy, 21 April 2000.

Examples of finishings and our final result are shown in **Photos 89–93**.

The Barn

This describes the basic conversion of a barn space into an open area (or in this case two, including its gin gan) with several potential uses, not requiring any plumbing works for kitchen or bathroom spaces and therefore acting principally as reception space. It is the least likely scenario for an amateur builder, but would apply to a house connecting to a barn or two separate barn areas.

Someone starting with a barn and intending to turn it into a complete dwelling rather than an additional open space for an existing house would combine this section with the Barn Conversion described later in order to create other rooms. Compared to the Farmhouse, the Barn was uncomplicated as we intended to leave the space open. The internal works were therefore straightforward. Building design and planning, build specifications and finding builders are described in the common works section, together with some of the building works.

The build itself

- Initial groundworks and services

These are described in the common works section.

Case studies – the builds in practice

- Masonry work, roofing and guttering

Other than the treatment of internal and external walls, the main masonry work involved in the Barn was laying a damp proof course as we planned to make no changes to the building fabric.

Pointing and the treatment of the inside of the external walls are described in the common works section.

The only other masonry work needed in the Barn was to install two padstones to support the ends of the beam spanning the Barn at the end of the mezzanine and to add a vertical post for additional support (at the request of the building control officer). The roof was in a good state of repair, just needing minor attention, and the only remaining work was to replace some of the wooden guttering, which had rotted, with the same reclaimed wooden guttering used in the Farmhouse.

- First-fix joinery

1. Floor joists and other supporting timbers

We only needed limited work for supporting joists and other beams, as the new upper floor consisted only of the mezzanine at the Farmhouse kitchen end of the Barn. To support the mezzanine we cut a reclaimed heavy 20x20mm pallet beam from one of a number of 6m long, approximately 20x20mm cross-section, darkish oak-coloured, long oriental hardwood beams of uncertain wood type, we had bought through an advertisement in the local free advertising paper for the equivalent of around £200 in today's money. These beams had come from a pallet, supporting what must have been an extremely large piece of machinery.

We found old reclaimed joists to support the flooring for the new mezzanine and floored this with the same Douglas Fir as elsewhere leaving the undersides of the joists and floorboards exposed in order to match the appearance of the underside of the mezzanine on the kitchen side.

2. Floors

We floored the whole of the Barn (and its gin gan) with the reclaimed tongue and groove Douglas Fir sawn from floor joists and pinned to 50x50mm battens fixed onto the concrete screed, using the same flooring for the two mezzanines. Laying the flooring was therefore straightforward.

There is a vast array of reclaimed floorboards available. These vary particularly in width, with costs tending to rise exponentially as this increases beyond around 15cm. Deepness of colour has a huge impact on the finished appearance and we were lucky to find high-quality, inexpensive, wide, tongue and groove Douglas Fir boards, which married a decent width (around 22cm) with a well-seasoned colour. We had used narrower boards (7cm) in the original Gin Gan and regretted it because of both the work involved in laying them and because they did not seem to marry as well with the old building.

Wood floors can be finished with any of a wide range of varnishes, waxes or other sealants. From our experiences in the early Gin Gan project, the only finish we found that protected the wood, remained resilient over time and brought out its colour was oil-based Bourne Seal®. This was one of the few tasks we physically did ourselves in the build, partly because it saved us several pounds/m² in sealing costs and partly to stay one step ahead of The Boys in order to protect the floors. However carefully these are covered with protective materials (which takes time) dirt inevitably penetrates the wood and the process of washing to remove dirt stains from clean wood is both laborious and never complete. We have not needed to recoat the floors over the 20 years we have been in the property.

This sealant also takes wax well and we have maintained the floors with a proprietary floor wax applied by an industrial floor-waxing machine, probably the only realistic option for some 150m² of flooring. The machine was donated to us by my hospital when they replaced their old stock. The reason for this became apparent once I started to use it: without care the machine will spin the user and not the polishing discs!

We found it best to finish other wood surfaces such as doors and architraves with wax. Solid wax is hard to apply and even harder to burnish if too much is applied, and we abandoned this for a semi-liquid floor wax (Luberon®) which seemed to work well.

We did flirt briefly with one of the liquid water-wax emulsions, which are easier to apply and burnish, but these did not actually leave much wax on the surfaces and rather defeated the object, and we had to recoat them with Luberon®. Waxing is a time-consuming process, but one which we did ourselves and saved a fair amount of a labourer's time.

A few examples of the internal barn works are shown in **Photos 95–98**.

3. Windows and doors

The Barn had two wall openings: one small window on the courtyard side of the building and a large, 3m wide x 4m high, opening on the other side. The gin gan had four openings between its pillars, differing slightly in width from 3–4m and a little over 2m high.

Case studies – the builds in practice

Photo 95 Barn. Attractive roofing A-frames and purlins. Thin, highly-insulating materials used to leave the purlins exposed.

Photo 96 Barn. Under construction: the reclaimed modern hardwood structural beams. Not old but fitted in well and cost pennies.

Photo 97 Barn. The reclaimed council radiator without its new cover and partly laid Douglas Fir flooring.

Photo 98 Barn. Finished Douglas Fir flooring (reclaimed old floor joists) imperfections fitted with overall look; sealed with Bourne Seal®.

We closed the smaller opening with a simple two-panel window and the other openings in the Barn and its gin gan above a 30cm high stone wall topped with cut, reclaimed stone flagstones over a damp proof membrane, with long vertical-glazed window panels and coated these with a black sealant to minimise visual impact.

Viewed from a distance these windows now appear as openings in the walls. Although we needed to use single-glazed windows in the Farmhouse because of the period window types, we opted for high-quality Pilkington K™-type double-glazing rather than triple-glazed alternatives in the Barn. Although the U-value of the triple-glazed windows was slightly better, the premium paid for the modest improvement seemed to be extremely high compared to the cost of adding extra insulation elsewhere in the building.

4. Staircases

The staircase in the smaller section of the original barn area was irredeemably damaged. We therefore removed it and replaced it with a new staircase, installed in the Barn space in order to access both mezzanines, one old, one new.

As described above, we constructed the staircase from wood cut from the same pallet beams we had used to form the support for the mezzanine. This was an oriental hardwood (type unknown) which, once waxed, produced a dark colour very similar to the existing roof purlins and A-frames.

The main issue which arose was the tensioned wiring we had proposed for the new staircase and balustrade which is now generally not approved, at least not for domestic buildings. It creates an open appearance compared to wooden spindles and we again used galvanised braided wire tensioned with turnbuckles in the same way as we had in the Mill.

This had been approved in the initial building control plans, but at the final building control assessment the building control officer was unhappy with the arrangement as the wires could be pulled apart – breaching the UK 100mm rule for gaps on staircases and balustrades. This posed a significant problem, as we wanted to keep the Barn as open as possible within the limitations of building control. Helpfully, the officer advised us that as long as we complied with building regulations at the time of his final 'sign-off' it was not his responsibility for anything we chose to do after the building had been approved; but obviously at our own responsibility. The problem was therefore resolved by fixing chicken wire inside the tensioned wire arrangement. What we did with said chicken wire later on we will leave to the reader's imagination!

5. Dry-lining

We did not install any dry-lining in the Barn as we left all of the external walls exposed.

6. Insulation and ceilings

If a space is to be left open, only limited work is needed to install floors, ceilings and insulation, heating, windows and a small amount of finishing joinery to turn an empty shell of a barn, for example, into an attractive and usable space. Ideally, the more open this is the better the final appearance, although most people will need to create rooms in the space. It is still, however, possible to do this and keep much of the appearance of an original barn (see the Barn Conversion Page 135, Photos 114–117).

We used multi-layered foil and foam insulation, similar to the current Superquilt® or Ecoquilt® types, rather than polyurethane in the Barn and its gin gan in order to obtain sufficient insulation to leave the maximum amount of the roof purlins and sides of the A-frames. This needed a degree of negotiation with building control as the company claimed from its own testing that one layer of their foil and foam insulation had equivalent insulating properties to around 150mm of fibreglass, the standard at the time. However, this claim relied partly on heat reflection in addition to insulation per se and it was not possible to establish an equivalent insulation U-value for the material with the regulatory test methods in use at the time.

I believe the modern-day products do comply with the regulations, although the reader should check with their local building control officer. This type of insulation certainly does minimise the thickness needed to insulate the roof of a building if structural timbers are to be left exposed.

The exposed external walls in the Barn slightly complicated the situation as these did not meet the necessary U-value for building control, but we could not cover them up because of the property's Listed buildings status (nor did we wish to). In reality, only around one-third of the total length of the Barn walls was actually external (the majority being either window spaces or walls separating internal parts of the buildings). We managed therefore to trade this off by over-insulating other roof areas in the property with double or triple thicknesses of polyurethane or foil/foam insulation.

7. Internal rooms

We left the spaces in both the Barn and its gin gan open, and separated the Barn part with no physical boundaries into a 'snug' at the kitchen end underneath the new mezzanine, a central dining area and a sitting area at the far end of the Barn where we installed a multi-fuel stove. We also intended the gin gan to be used as an additional sitting area. Together, this gave a disproportionate amount of 'reception' space in the overall Farmhouse and Barn – and too much space in general – which was one of the reasons why we decided to carry out the later Barn Conversion.

- Plumbing, heating, electrics, renewables and other secondary heat sources

The heating requirements of the system serving the Barn and gin gan were rated as 30–35kW, and in order to provide the necessary heat we would have needed a total length of around 10m of conventional 600mm high double-panel convector radiators, which would have occupied a large amount of the available wall space. We managed to find a number of reclaimed high-output former council building radiators, bought for a few pounds: these are approximately 1–1.5m long, 500mm high and 250mm deep and are fed by 25mm copper pipe into wide-bore internal pipework, enabling a far greater water flow than a conventional radiator. This system is combined with an integral high-output fan. One 1.5m long radiator has an output of approximately 10kW so, combined, they hugely reduce the wall space needed for radiators and deliver a large amount of heat, quickly, because of the fan. We found these to be invaluable in our large, potentially cold, open space. Despite having been well used when we bought them, they have been robust and have functioned almost faultlessly for over 20 years. It is still possible to find similar models.

When we initially converted the Barn, we also bought something new for a change: a large Coalbrookdale Darby multi-fuel stove, a bigger version than the Severn, with a total potential output, fire and back boiler together, of up to 16kW. We plumbed this into one of the ex-council radiators and found that this combination could comfortably warm the Barn on a cold day. The Darby, however, was perhaps a step too far, as the size of the firebox means that it needs to be stoked very regularly and is prone to smoking if it does not achieve a rapid burn when lit, as it needs a wide-bore flue (rather than the more typical 125–150mm diameter). The smaller firebox and narrower-bore flue of a smaller fire appear to make the fire more manageable, both in terms of lighting and keeping alight, at the cost of only a slightly lower overall potential heat output. As with the Farmhouse, using a multi-fuel stove with a back boiler in the Barn combined with a high output air-source heat pump has hugely reduced both our oil use and carbon footprint.

- Finishing joinery, other finishing works, decoration and finishing groundworks

There was little finishing joinery in the Barn, as the space was open-plan and the walls were exposed and therefore unsuitable for skirting boards. The only additional joinery

Case studies – the builds in practice

work we needed was to scroll the floorboards into the wall (another reason for not using new boards; older ones are far more visually forgiving).

The Barn had no kitchen or bathrooms and, apart from the exposed walls, the only decoration needed was emulsion paint to the ceilings. The finishing groundworks are described in the common works section.

Photos 99–104 show some finished appearances of the Barn building.

The Byre

This is an example of a simple way of turning an outbuilding into a useful independent, additional living space. At the time we intended to use it as a nanny flat, but it is now used for letting income.

Internally, and not unlike the earlier Garage project, we used the space to create two main rooms – a bedroom and an open-plan kitchen, dining and living area, separated by simple stud partition walls – and added a couple of additional partitions to close the area between the Barn and the Byre. This left a narrow space between

Photo 99 Barn. Finished Barn looking away from the Farmhouse. Open-plan 'snug' with the dining and seating areas.

Photo 100 Barn. Finished Barn looking towards Farmhouse with reclaimed Douglas Fir flooring and exposed roof timbers.

Photo 101 Barn. The finished Barn: view from the new mezzanine.

Old Buildings: Conversion and Restoration

Photo 103 Barn. Finished view of the gin gan. Left much alone other than adding a floor, windows and ceiling boards.

Photo 102 Barn. The arrangement of the gin gan timbers is similar to the kiln end of the Mill: similar pyramidal roof shapes.

Photo 104 Barn. Outside north side: minimal change from the 1964 aerial photograph (Photo 18) or early appearance (Photo 22).

the two, sufficient for the plumbing hub and electricity consumer unit for the Byre and the boiler serving both the Barn and the Byre, with the remainder kept as a storage area. We created a flat-ceilinged bathroom in the large Byre space in the opening between the bedroom and the open-plan kitchen, dining and living area. We also kept two of the 1m high animal stall separators in the bedroom as they were perfectly spaced to accommodate a double bed and bedside tables, together with a stall separating the kitchen from the living area. Together these preserved at least part of the history of the building as a cow byre.

Although architecturally not particularly old and with few period features, because of the attractive Victorian purlins and A-frames and the overall style of the building we left the ceilings vaulted, leaving the crossbeams of the old pine A-frames exposed. We could not increase the usable space in the building by adding an upper floor because of the relatively low roof height and the width of the building.

Building design and planning, build specifications, finding builders

These are described in the common works section.

The build itself

- Initial groundworks and services

These are described in the common works section.

- Masonry work, roofing and guttering

We only needed minimal masonry work, other than to point the external wall and expose an internal wall, install a damp proof membrane, cover this with a concrete screed and remove some of the concrete animal stall dividers. We did not create any new wall openings. Unlike the rest of the buildings making up the farm complex, the Byre roof was in poor condition and we decided to replace this as repair carried too much risk. The roof area involved was relatively modest (around 100m²) and the vast majority of the pantiles themselves as well as the main roof structure were intact. We therefore only needed to buy roof battens (cheap) and find several square metres of old matching pantiles, which we achieved courtesy of a local reclamation yard.

We were able to reuse the existing cast iron guttering.

- First-fix joinery
1. Floor joists and other supporting timbers

As the Byre was single storey, we only needed 50x50mm battens to fix the flooring above the damp proof membrane.

2. Floors

These were modern whitewood, which we carpeted.

3. Windows and doors

Our local fabricator made these as identical replicas of the original windows. We managed to find three reclaimed Victorian doors to fit the openings to the bathroom, bedroom and small internal boiler house and storage area, and a pair of simple glazed external double doors, which added considerable additional light.

4. Staircases

We did not require any staircases as we could not create an upper floor because of the height of the A-frame crossbeams.

5. Dry-lining

After some discussion, the building control officer agreed that we could leave one internal side of an external wall (the warmer, west-facing wall) and the short section of south-facing wall exposed, dry-line the colder, east-facing wall and add additional polyurethane insulation to the roof, which we achieved without covering the purlins. We used the same dry-lining technique with a polythene membrane as in the other parts of the farm complex.

6. Insulation

We left the inside roof open to the eves and insulated this with a double layer of the reclaimed polyurethane used for the rest of the farm complex.

7. Internal walls

We built the few internal walls needed using simple insulated stud partitions to create the bedroom and bathroom spaces.

- Plumbing, heating, electrics, renewables and other secondary heat sources

Once we had installed the water and electricity supplies in the Byre, we ran the necessary pipework and cabling underneath the floor to the various areas in the building. This was uncomplicated. As we had installed a boiler to supply heating to the Barn, we added a zone valve to the heating system with a separate thermostat to isolate the Byre from the Barn.

As in the rest of the farm complex, we did not install any renewables at the time, although five years later, once these had come into more widespread use, we installed a 4.5kW air-source heat pump in the open-plan kitchen, living and dining area in the Byre which supplies a large proportion of the heat required. We did not install a multi-fuel stove because of planning restrictions.

- Finishing joinery, other finishing works, decoration and finishing groundworks

As we converted the Byre into a modern living area inside the original cow byre, all of the finishing works were relatively straightforward.

We used stained whitewood for door architraves and skirting boards and reclaimed Victorian four-panel internal pine doors for the bathroom and bedroom. We found the door furniture either from items still in the original buildings or from various markets and junk shops. The kitchen units were reclaimed and the bathroom suite was new and simple, although we did restore an old Victorian high-level cast iron cistern from the original Victorian end of the Farmhouse – this was sandblasted and treated externally with wax and turpentine as it was a rather interesting piece of, probably late Victorian, cast iron metalwork and has lasted the 20 years we have been here.

Finishing groundworks are described in the common works section.

Some before and after examples of the Byre build are shown in **Photos 105–108**.

Case studies – the builds in practice

Photo 105 Byre The outside of the original Byre facing into the courtyard.

Photo 106 Byre. The same wall after cleaning and repointing, with replica windows matching the damaged originals.

Photo 107 Byre. A 'bait break'. The frame for the bathroom has been erected and the animal stalls are still in place, some dividers to be kept.

Photo 108 Byre. Finished. Shows roof timbers, living and kitchen area (demarcated by animal stall divider), bathroom and door to bedroom.

The later Farmhouse Extension

We carried this out five years after the initial works. It merits a brief description, principally because of the issues relating to planning permission and a couple of relevant illustrations of aspects of the build process itself. We designed the Farmhouse Extension more from the perspective of being sympathetic to the buildings rather than to save money, although we did reduce the total potential build costs significantly for the reasons explained above. Being south-facing it makes optimal use of sunlight and is warm most of the time. It is a good example of where we used an air-source heat pump because of difficult access to the central heating system in addition to significantly reducing both our running costs and carbon footprint. We used the early aerial view (Photo 18) shown alongside the description of the farm complex in Chapter 3 to support our planning permission and Listed building consent application as it shows the earlier corrugated-roof structure extending into the courtyard.

In terms of cost, in most cases this type of build is likely to attract VAT at the standard rate so the cost-benefit of using individual tradespeople who are not VAT-registered is considerable. This amounted to around 18% of our total potential build costs, as we needed few materials; the VAT payable was further reduced as we bought the flooring from someone who was not VAT-registered as a leftover from his own build.

Despite the overall space available in the Farmhouse and Barn together, the space in the open-plan Barn areas was realistically too large for everyday use.

Some 10 years after the original restoration, we therefore decided to apply for permission to add an additional south-facing and mostly glazed extension to the Farmhouse – leading out from the kitchen and into the courtyard.

Building design and planning, build specifications and finding builders

We needed planning permission and Listed building consent in view of the Listed status of the farm complex, although this would not normally be necessary for an extension of this size as it would be within permitted development limits (providing no earlier extensions have been made to the property).

The design was straightforward as the addition was a simple rectangular room, mostly glazed above a 500mm wall, with a corrugated Onduline®-type roof. We added this design to the plans the architect had drawn up for the original farm complex builds, which cost next to nothing as all of the existing plans and elevations were all available and had been paid for previously. As usual, these included the basic building control requirements.

We obtained permission for this probably, at least in part, because we had historical evidence that a similar, corrugated-roofed structure had existed from a 1964 aerial photograph (see Photo 19, Page 62).

Our intention was to build an extension that was as close as possible to the original but still create a functionally useful space. We continued to work with the Boys for this build.

The build itself

- Initial groundworks and services

We did not need any groundworks apart from digging trenches for the foundations.

- Masonry works, roofing, guttering, and first-fix joinery

As these were simple, I have not broken them down in detail into their component headings. We designed the room to be approximately 6m x 3.5m, with a build structure consisting of an insulated cavity wall with stone on the outside and blockwork internally, high vertical panelled windows made by a

Case studies – the builds in practice

local fabricator and a gently sloping corrugated roof with cast iron guttering emptying into the existing water drains. We left the area of previous external wall stonework, now internal, exposed.

We intended our roofing to match the previous structure in terms of type and colour, for appearance rather than cost. As the aerial photograph showed a brick red-coloured corrugated panel roof, we created a gently sloping roof structure using Onduline®-type sheeting over a gently-sloping ceiling, which we insulated with 25cm of fibreglass. This decision did, however, reduce our costs by a few thousand pounds through the combination of the lighter roof structure and corrugated sheeting instead of pantiles. We also considerably reduced our flooring costs by using reclaimed pre-sealed Junckers® engineered beech boards, which were not liable for VAT (as described above).

- Plumbing, heating, electrics, renewables and other secondary heat sources

We decided to install an air-source heat pump as the room was out of immediate reach of the central heating system without major disruption. In addition to avoiding the cost and need to lift significant amounts of flooring to connect to the existing heating system, this decision had a major advantage in terms of future running costs and our carbon footprint given the heavy use of the room. It also avoided the cost of unnecessarily heating rooms which we did not use during the daytime (see further discussion in the Overview, Chapter 5).

- Finishing joinery, other finishing works, decoration and finishing groundworks

We added some low-level internal panelling to one wall, with an internal pine windowsill running the full length of the wall. As the internal ceiling was flat we simply embedded LED spotlights. Other than this, the room did not require any finishing works.

Finished views are shown in **Photos 109–112**.

The later Barn Conversion

We carried this out 15 years after the initial build. It was a further project on the main Barn which we had earlier converted to provide an open-plan living space to change this into a new independent dwelling with kitchen, bathroom and (three) bedrooms included in the original space.

This project is an example of redeveloping a large open space to create a modern dwelling yet retaining the openness of an original barn. We carried out the Barn Conversion

Old Buildings: Conversion and Restoration

Photo 109 Farmhouse Extension (right). Closed-up redundant entrance, outbuilding doors and windows painted heritage blue. In keeping with the original, the extension windows and entrance doors are finished in antique pine (Sadolin®).

Photo 110 Farmhouse Extension. Front view of the Extension.

Photo 111 Farmhouse Extension. The corrugated roofing blends in colour with the Barn roof as in the early aerial photograph but, as it was originally, the roofing type is quite different to the main building. Not done to save money, but this construction was also considerably less expensive.

Photo 112 Farmhouse Extension. Internal view of the Farmhouse Extension. As it is south-facing, the room attracts heat and is warm even on cold sunny days. It has become one of the most used rooms in the Farmhouse.

because the large original barn area had become a glorified entertaining room used only occasionally, and we therefore applied for permission to convert it into a separate property.

We combined an 'old and new' approach, creating a modern conversion but keeping all of the features of the original barn including the exposed walls, old Douglas Fir flooring, vaulted ceiling with its purlins and A-frames and the original supporting beams for the milling equipment in the gin gan. With these period features spread throughout the building we needed a delicate approach to avoid the new appearance clashing with, or distracting from, the old. Our aim was also to preserve the appearance of a barn as much as possible.

Building design, planning, build specifications and finding builders

These are grouped together as all were straightforward. The main part of the project began with the planning research stage. We resolved this with a couple of telephone calls to the planning department and submitting an approximate drawing and broad proposal for initial pre-planning advice. This received in-principle approval, probably as a result of some relaxation in the planning regulations since we had carried out the original three farm projects.

We were perhaps fortunate in the design of the Barn Conversion, as the Durham City senior conservation planning officer with whom we had dealt very successfully for the original farm projects was, 15 years later, working as a private planning consultant. Chris was hugely helpful in assisting us with the design and layout of the new converted barn space. We chose to create two bedrooms, a single bathroom, a large open-plan dining kitchen with a seating area and the gin gan as the main sitting room on the lower floor and a further bedroom created from the original Barn mezzanine combined with a new mezzanine on the upper floor overlooking the remaining half of the Barn. Whilst we could have created some additional accommodation by using more of the upper space, this would have been at the cost of a third staircase and the loss of the overall appearance of a barn. In order to create rooms within the available space and to obtain sufficient natural light our proposal included installing two new wall openings and an additional rooflight. As we had already obtained permission for two rooflights in the Barn during the original build, the planning department did not object in principle to the addition of a third. In terms of other openings, we were also fortunate in finding two additional openings in the original barn which had obviously been blocked up at some stage in the life of the building. The stonework in these openings clearly demarcated their boundaries and the original wooden lintels were still present. We were therefore able to provide clear evidence of the existence of earlier openings to the planning department; this was a major factor in obtaining permission.

The external structural changes involved installing a second external door in one of the blocked up wall openings and a window in the second. The internal structural changes involved closing the two internal access doors from the upper and lower original smaller areas of the barn (now the Farmhouse kitchen) and its upper mezzanine to separate them from the Barn, which we combined with fireproof and sound-proofing insulation. We also added a new mezzanine with a second access staircase in the former barn area; this was needed because of the height of the crossbeam of the A-frame, which did not allow access across the upper floor. We moved the large multi-fuel stove from the original main Barn area into the gin gan but sacrificed its back boiler as the gin gan had no roof space for the necessary plumbing. This needed a small amount of masonry work to dismantle the existing fireplace, lay a new flagstone hearth and erect a steel chimney in the gin gan. The only other structural change involved closing the previous access to the Byre, again combined with fireproof and sound-proofed insulation.

We agreed the detailed build specifications with The Boys as we went along, working from the building control plans rather than defining the specifications in advance.

The build itself

- Initial groundworks and services

Water and electrics supplies were already present in the original Barn, although we separated the mains supply to the Farmhouse to make the two supplies independent. The only other groundworks involved adding the second oil tank and digging a trench from this second tank to run oil to the building.

- Masonry works, roofing and guttering

We only needed limited masonry work to create the openings described earlier and to support the wall above the new window opening, leaving the original external and internal wooden lintels in place.

As described above, we closed the two openings into the Farmhouse and added the fireproof and sound-proofing insulation to the roof apex, keeping the original doors on the Farmhouse side as façades. The brickwork did not stand out particularly on the Barn side and we therefore decided not to follow our original intention of panelling over this. Old barns are often full of a mix of stone walls with brick patching of closed-up sections, so an additional couple of bricked up openings did not look out of place (though we chose the brick type carefully). We left all of the original internal walls exposed apart from a small façade directly behind kitchen wall units, which needed to be a flat, flush surface for the wall units.

We did not need any roofing or

guttering works apart from installing the additional rooflight, which we were lucky enough to find secondhand online (as described earlier, conservation rooflights are exceedingly expensive).

- First-fix joinery

The majority of the first-fix works were uncomplicated, but formed the major part of the Barn Conversion. These mostly involved constructing the stud partition walls on the lower floor to create the two bedrooms, bathroom and an oblique corridor leading to the existing staircase up to the original Barn mezzanine, which formed the third bedroom. We created the new mezzanine and installed the second access stair from the remaining large Barn area. The upper room and mezzanine took up around half the length of the original Barn, leaving the remaining section, a lower open space of approximately 8m x 6m, open to a vaulted ceiling. We also left both of the upper areas (the former mezzanine, now a bedroom, and the newly created mezzanine) open to the roof apex. All of the roofing space had already been insulated. The total finished floor area was approximately 190m².

1. Floor joists and other supporting timbers

The only timbers which we needed to install were the supporting beam and the floor joists for the new mezzanine. This was straightforward. We still had two of the long, thick hardwood pallet beams from the original work 15 years previously and used one to span the width of the building as the support and also added an additional vertical post, cut and set into a steel shoe fixed to the concrete floor screed in order to add extra support for the crossbeam.

2. Floors

We kept all of the original Douglas Fir flooring in the original Barn and used reclaimed Junckers® engineered beech for the flooring in the new mezzanine. As this area was separate from all of the other rooms in the barn and only had a limited amount of exposed internal wall area, the modern beech did not seem to clash with the existing features.

3. Windows and external doors

We had the one new window fabricated locally: this consisted simply of two vertical double-glazed panels, and we sourced a reclaimed, glazed external door).

4. Staircases

As described in the design section above, we needed a second staircase because of the height of the crossbeam part of the A-frames. This was constructed by a local fabricator from Paraná pine which, although new wood, is attractive and relatively inexpensive. When varnished it left a warm colour which was not a match to the original Douglas Fir boards, but sat comfortably alongside them.

5. Dry-lining, insulation and internal walls

We left the walls exposed and therefore did not need any dry-lining. Likewise, as we had already insulated the Barn, the only insulation added was in the new internal walls and for the ceilings between the new lower rooms and the upper floor. The internal walls were constructed using insulated plasterboard for the stud partitions with a finishing skim of plaster.

- Plumbing, heating, electrics, renewables and other secondary heat sources

All of the additional plumbing works (copper piping this time and not plastic!) had been installed when the new services were connected. We added to these, running the necessary pipework underfloor to the kitchen area and bathroom. As the original build had two boilers, in order to separate the Barn from the Byre, we needed to change this plumbing arrangement and install a third boiler so that the Barn, Byre and Farmhouse each had their own completely independent central heating systems. Again, fate was kind to us as we were able to obtain both an oil tank and a condensing boiler secondhand from our farming neighbour who had just installed a biomass boiler, knocking several thousand pounds off the total Barn Conversion cost. Installing the boiler, separating the plumbing systems and adding the gravity-fed system for the new boiler were all relatively straightforward and inexpensive. The Barn Conversion therefore had three heat sources, the oil central heating system, two air-source heat pumps (see below) and the large Coalbrookdale multi-fuel stove.

As the large air-source heat pump in the original barn had been very successful, we added a second, 4.5kW output, heat pump, this time in the gin gan. This pump, together with the heat pump in the main open-plan area, provided approximately 12kW of heat output, consuming approximately 3.5kW of electricity when running at maximum output. The fire and heat pumps together can heat the building in all but the coldest of weather.

- Finishing joinery, other finishing works, decoration and finishing groundworks

For the second staircase and mezzanine, instead of using

conventional wooden spindles we bought toughened glass panels, approximately 1m² each, for the mezzanine balcony, together with shaped pieces to line the staircase itself; these cost approximately £60/m² panel, similar to the cost of buying wooden spindles and accessory parts, but avoiding a huge amount of joinery work. They also had the great advantage of being completely transparent and left an open appearance throughout the staircase and upper floor of the Barn Conversion. We combined these with a large panelled internal window in the upstairs bedroom, overlooking the new mezzanine and lower barn area.

We used painted whitewood skirting boards for the stud partitions and facing architraves for the doors, although the doors themselves were modern oak for the downstairs rooms and entrance to the staircase to the upper room, again bought online. We also obtained an opaque glass-panelled door for the bathroom to add more natural light into the short corridor and a similar oak door with a modern-shaped glass panel between the new kitchen/dining area and the gin gan to replace the original solid ledge and brace door, again to maximise natural light. Although the doors were modern, the combination of old and new did seem to work in overall appearance.

For the kitchen, we installed inexpensive modern units (gloss white, kept as simple as possible, and bought from an online supplier) and engineered beech worktops, which fitted well with the exposed stone walls and did not draw attention away from the other features in the Barn. We used the same simple approach for the bathroom fittings.

Decoration was equally simple – plain white. We had evolved from various colours and types of paint in the Farmhouse and concluded that white or off-white worked best with the old walls and other period features. Obviously, this is a matter of personal taste. We did not need any finishing groundworks.

This conversion is illustrated in **Photos 113–122.**

Old Buildings: Conversion and Restoration

Photo 114 Barn Conversion. New mezzanine, view towards Farmhouse. Glass panels open up the building avoiding wooden spindles.

Photo 113 Barn Conversion. Old closure in a wall used to support the planning application.

Photo 115 Barn Conversion. Panoramic view from the mezzanine over the open-plan area.

Photo 116 Barn Conversion. Same area, view away from Farmhouse with new entrance.

Photo 117 Barn Conversion. Towards the Farmhouse: open view through the building.

Photo 118 Barn Conversion. Little changed in the gin gan other than to relocate a stove.

Case studies – the builds in practice

Photo 120 Barn Conversion. A lower bedroom, same wall opening and floor and the closed up entrance through to Farmhouse.

Photo 119 Barn Conversion. A slightly risky choice of door to add natural light in an old building but it seemed to work.

Photo 121 Barn Conversion. The difference in quality and appearance did not seem to justify the high cost of specialist suppliers.

Photo 122 Barn and Barn Conversion. Old fan-assisted radiators. Covers made on site: rapid heat and saved wall space.

Photo 123 Barn Conversion. Destined to become a family heirloom. Another find along the quest for materials.

Chapter 5: Overview

This book describes some principles, ideas and solutions we used together with our practical experiences from a range of restoration and conversion projects. These began with some simple upgrading of properties in Oxford and Glasgow and some finishing works in a contracted project in County Durham, the Gin Gan, which I have not described in any great detail. The main descriptions begin with a small project, a modern garage conversion, and continue with six larger projects managed completely by ourselves: a derelict mill, a period farmhouse, its barn and gin gan, a cow byre, an extension to the renovated original farmhouse building and the further conversion of the original barn to create a separate dwelling. The descriptions illustrate the issues we were faced with and the decisions we took in order to achieve both the desired final result and preserve the heritage of the buildings, while working on a limited budget.

The options we chose are by no means a recommendation to abandon the more conventional approach of appointing an architect to design, draw up plans, specify the works and oversee a project but are simply examples of alternatives. For some buildings an architect may be invaluable, if not essential, to provide design options and possibly to manage a building project. In many of ours the designs were either very simple and/or were dictated by the fabric of the buildings themselves (rooms and wall openings) and we wanted to project manage ourselves, but to establish a model where we could obtain specialist advice when we needed it. In these cases, we mostly limited our use of an architect to turning our designs into detailed plans for submission. Some of the decisions we took were not conventional and a few will be criticised by experts. They have all, however, lasted the course of time. Someone setting out to convert or restore a building could use anything from none to some or all of the approaches described.

The decisions we made are summarised in approximately the same sections as they appear in Chapter 2, excluding the issue of grants which is discussed and dismissed in Chapters 2 and 4.

Subjects examined:
- Risks and rewards
- UK Value Added Tax (VAT)
- Building design, plans and planning
- Build specifications
- Finding and instructing builders
- Old versus new and repair versus replace
- Materials: new and reclaimed
- Fabrication joiners
- Old versus new and repair versus replace
- Heating, renewables and other secondary heat sources
- Maintenance and running costs: impact of build solutions
- Modern versus old approaches to the buildings

Risks and Rewards

We discussed the specifications for the build options with an architect and surveyor in the case of one project (the Mill), with our builders in the three main projects on the farm complex buildings and the Farmhouse Extension and with a planning consultant in the Barn Conversion. In most situations, and particularly with the builders themselves, we were often given three options to consider in terms of cost and risk. Usually, the least expensive option carried risks of not lasting over time or potentially costing more in the longer term. The decisions described in the case studies were usually the second of the options and were probably mid-range in terms of risk. We avoided 'cheap' solutions but equally found that taking a small amount of risk in many situations reduced costs enormously compared to the most expensive 'gold-plated' solutions. In some cases, such as retaining and repairing some original structures (for example not reroofing most parts of the farm complex, and repairing flooring and a staircase in the Farmhouse) we reduced costs considerably and also helped to avoid changing the original appearance of the buildings. These decisions were also supported at the time by the conservation planning officers. At least in our own opinion and that of our conservation planning officers, our decisions also gave a better final appearance and were the most faithful to the original building. With the benefit of 20 years' experience, the decisions we took have stood the test of time and we have needed no more than routine maintenance.

UK Value Added Tax (VAT)

It is difficult to estimate the potential VAT liability on a project and users will need to take advice on what qualifies for full VAT relief or the lower rate tax and follow the rules to the letter. In the UK, some degree of VAT relief is likely on any old building which has not been inhabited recently, at least on a large proportion of the works (excluding, notably, expert opinions). Based on current rules and rates we, in practice, reduced our own VAT liability to close to zero, saving around 5% of total costs by using individual tradespeople who were not VAT-registered compared to larger contractors on most of the projects (those on which the lower 5% rate would have applied). The potential saving however can be anywhere up to a little short of 20% of build costs if the building does not qualify for beneficial VAT treatment (in our case the Farmhouse Extension and later Barn Conversion).

In terms of VAT and depending on the rules that apply to the build, contracting individual tradespeople who normally work on a 'fit only' basis rather than 'supply and fit' may also help to reduce their own need for VAT registration and therefore the amount of VAT payable. Some tradespeople prefer to work this way as it means only keeping the necessary records for income tax purposes, avoiding the burden of keeping VAT records, though this might change in the future with 'Making Tax Digital'.

Building design, plans and planning

Using our own design or enlisting support from a planning consultant and limiting the input of an architect to drawing up the plans and elevations required in several projects reduced the overall cost of fees by around 15% compared to using an architect for the whole process at no discernible detriment as the plans were simple. In retrospect, the design for the Mill was also straightforward and we could, with more experience, have used our own thoughts for the design stage (and the later specifications), although it was a mistake not to have appointed a project manager when we were trying to run the project from several hundred miles away.

Build specifications

Our decisions to reduce some works or use alternative building techniques were not all conventional. All, however, saved a vast amount of money (10% or more of the total potential build costs for pointing the stonework alone). In addition, over time our decisions have not led to any obvious problems (such as damp, condensation or any evidence of damage to the stonework from our decision over how we pointed walls). Using mid-priced or mid-risk solutions combined with advice from experts or from our builders themselves had a huge effect both on reducing costs and the final results.

Finding and working with builders

Regardless of the decision on whether to work with a larger contractor or individual tradespeople, personal recommendations are almost certainly the most robust way of finding the 'right' company or, particularly, individual tradespeople. We were very fortunate in knowing excellent builders from one of the initial projects (contracted works on the Gin Gan) who could also recommend specialist tradespeople when required for the farm projects, but we knew few people located near the first large project (the Mill) and did not, therefore, have personal recommendations available. We did not think at the time of the possibility of asking a builder local to where we were living to 'relocate' in order to carry out the project 300 miles away. In this case, we chose tradespeople from a listing of local services available at a time when the internet was not as mature and online reviews were not available. After running into difficulty following the masonry part of the Mill build, we turned to our local (and only) building supplies merchant – a rather obvious route to take but one we had not thought about at the start! From there, we found a dedicated joiner, possibly a more appropriate person to carry on the majority of the later works than our initial builder, who was primarily a stonemason. There was clearly, therefore, some risk in the decision we took when we started the Mill works. This, combined with our failure to appoint a project manager, led to avoidable extra costs and a lot of anxiety before we handed the work over to the experienced joiner, who came with at least one local recommendation. Despite this, we still made considerable savings compared to using a large contractor from the quotations which were returned for our architect's tender specifications. Our new joiner was able to not only remedy the failings of his predecessor at a relatively low cost, but also to project manage the other tradespeople who became involved in the project. On balance, therefore, we have no regrets about using small operators as the only failing resulting from this decision was probably of our own making at the time. We found that paying our builders on an hourly basis was, contrary to received wisdom, a successful approach.

Small versus large

In addition to potential VAT liability, individual tradespeople will generally cost (very approximately) 25–30% less than larger companies. This carries some risk compared to using large better-known companies, hence the need to research tradespeople extremely carefully and look for personal recommendations. Many experts will prefer, however, to use larger contractors who can provide warranties on the work performed. This is a balanced risk but one which suited us well.

Old versus new and repair versus replace

The decision to repair versus replace roofs, flooring, staircases and other existing joinery saved up to 20% of total potential build costs, if complete reroofing of the various buildings is included, and around 5–10% if not. Excluding reroofing, where our decision just to repair obvious defects in the existing roofs carried some risk, our decisions to repair rather than replace were relatively uncontroversial, although many experts would simply have recommended replacement. Repairing certainly produced a closer reproduction of what had been there originally. Some will prefer to replace rather than repair, although it may not always be possible to find the necessary reclaimed materials, and new materials may be both extremely expensive and can potentially clash with the old appearance of the building. The use of purely new materials throughout a project, however, is a matter of personal taste and will depend on whether the owner wants a new interior finish or prefers a more faithful reproduction of the original.

Materials: new and reclaimed

There is, obviously, nothing controversial about using new materials assuming the owner takes advice from an expert or from the builders themselves about what to choose. Like the building solutions, the range of new materials available can be classified simplistically as the gold standard, the midway and the cheap and cheerful. Again, and rather like the building solutions, the modest benefit of gold-standard materials compared to the midway options may not always justify the large difference in cost.

In terms of reclaimed materials, in most cases the options available are not particularly controversial and few carry any significant risk. Some, such as using reclaimed (or refurbished) materials such as boilers are high-risk decisions and ones we would generally have avoided, although we were fortunate in finding some items, including boilers, locally from a trusted source. It is useful to extend the search for reclaimed materials far beyond the obvious such as flooring as in many cases such as joinery (doors) and kitchen/bathroom fittings there is a limited secondhand market and reclaimed items can be bought for only 10–20% of the new cost. Clearly this restricts choice to what is available but depending on the use (such as for a let property or a nanny flat etc. personal preference ceases to be relevant and the only requirements are quality and compatibility with the inside finish of the building. There are many sources for these, including local reclamation yards and online sellers (e.g. eBay, Facebook Marketplace, Gumtree, Preloved and local websites etc.). There are also many relatively low-cost transport services available. Added together, a careful search for reclaimed materials throughout the build can achieve many smaller potential savings and can still have a large impact on final costs. Overall, we saved over a third of the total potential materials cost because we did not restrict ourselves to conventional new sources alone or the most obvious reclaimed items.

Some of the materials, notably insulation for all three initial farm projects and a boiler and oil tank for the later Barn Conversion, were one-off finds. The opportunities for real bargains in reclaimed materials are probably less now than it was 20 years ago as the true value of these materials is more widely recognised. Other items, such as our one-off find of reclaimed high-output council radiators for the Barn (see Photo 123, Page 135), did not have a huge impact on cost as radiators are not expensive, but did have a large impact on the final appearance and left us with considerably more unoccupied wall space avoiding the appearance of walls lined with radiators in this example. For any would-be amateur builder, therefore, it is still well worth remembering the adage 'seek and you shall find'.

Fabrication joiners

We used fabrication joiners (from personal recommendations) in all of the projects rather than approaching specialist companies to provide, for example, windows (potentially the largest cost of joinery materials), but also to construct staircases, create boards from reclaimed materials for doors, door jambs, architraves and a range of other finishing joinery woodwork. We cannot compare the costs of our fabricator to those of specialist companies as we did not obtain formal quotations from these, but from our limited research, specialist companies were several times more expensive, and the items produced by our local fabricator were of first-rate quality and identical replicas of the originals.

Heating, renewables and other secondary heat sources

As we used renewable energy sources and other secondary heat sources (multi-fuel stoves) in conjunction with each other to great effect, I have combined them in this discussion. Renewables in our projects were investments rather than 'cost-effective' aspects of the builds themselves as they did not replace but complemented conventional oil-fired central heating systems. Excluding wind and solar energy, which we decided were not cost-effective in our projects and are discussed in Chapter 2 (Page 39), the solutions we chose have had an enormous impact on reducing our heating costs and carbon footprint.

Other renewables, such as ground-source or air-to-water systems and biomass boilers, were either impractical in our old buildings or were not cost-effective in our case, excluding government incentives which were limited at the time we installed them. The single renewables solution we chose was the air-to-air heat pump. We found these systems to be highly cost-effective, with a payback time of a year or so (or even less compared to our use of an oil-fired Rayburn stove). They were not disruptive to install and were relatively inexpensive (the four systems installed added around 3% to the total final build costs for the whole farm complex of buildings). Although some experts would disagree with our conclusions about renewables, we felt that many were simply not cost-effective and only generated limited amounts of power (electricity). At the time, the incentives available did not change this conclusion. Our decision to use air-to-air heat pumps is not controversial and their benefits were apparent 15 years ago, long before they started to feature regularly in the general media.

Like renewables, our other secondary heat sources (multi-fuel stoves, burning mostly wood) were also additional build costs, or investments, and have also greatly helped to reduce both our running costs and our use of fossil fuel. They are also attractive and many would install a stove anyway as a feature in a room. Regardless of the reason why one is installed, it would be wise to consider a model with a back boiler strategically positioned in a house (or barn). The plumbing required is not difficult if the necessary space is available and this type of system generates considerable additional carbon-neutral heat. The systems are not expensive to install (in our case around 1–2% of the total build costs for three stoves, two with back boilers at the time of the original builds). They provide almost immediate heat and together with the radiators plumbed into the back boiler, can heat the core of a house without the need to use the central heating system itself.

Various combinations of renewables and secondary heat sources can be used to reduce the use of an oil or gas-based central heating system. Some examples are described below. Our decision to install these is not controversial and carried very little risk: our one secondhand stove has functioned faultlessly for 20 years.

Impact of where the various heat sources are located

A large amount of information is available in the media and online about types of heating methods (heating systems, fuel types, boiler types, radiator types, thermostatic radiator valves (TRVs), zone valves and renewable energy sources). Less, however, is written about the location of heat sources and the impact this can have on the use of a fossil fuel-driven central heating system. Assuming that a restored house or converted barn has some form of central heating system, it is a truism to say that fuel will only be consumed if this system is running. Apart from the position of the thermostat (or thermostats) and any zone valves used to isolate plumbing for the heating in different areas of the house, the positioning of additional heat sources and how rooms are used during the day has a huge impact on their efficiency in terms of reducing central heating requirements.

Conventional boiler-driven systems will typically be programmed to come on at set times during the day, possibly with different time settings for space and water heating. A standard gas or oil-based central heating system can be optimised using a combination of TRVs and zone valves (typically isolating the living and bedroom areas). We did not do this in the Mill (but could and probably should have). With the configuration of rooms in the Farmhouse, in which we had bedrooms on both floors, it would have been difficult to use a zone valve, but we could and did install one in the Barn. Even if a zone valve is used and programmed effectively, it is inevitable that when the central heating system is running it will heat at least some parts of the house which are not in regular use. Secondary heat sources can circumvent this.

For anyone reticent about the

prospect of lighting a fire every day, things have changed a lot since the days of half an hour with expensive kindling and newspaper. It is easy to incorporate lighting a fire into the morning routine: in cold weather I sit by the fire with a morning coffee and pop a couple of inexpensive broken compressed heat logs (we use Hotmax® very effectively) and a firelighter into the kitchen fire. Two minutes and it's away! The sitting room fire takes over as our secondary heat source in the evening and provides heat from the back boiler to radiators in two bedrooms until all are safely tucked up in bed.

The effect of the secondary heat sources became most apparent after we replaced a continuous oil-burning Rayburn with a multi-fuel stove in the kitchen area. Apart from the savings in fossil fuel burned to heat the space in question, we were able to better optimise the amount of time during which the central heating system was used when this set of rooms was adequately served by secondary sources (see below).

Benefits of using combined heat pumps and stoves either alone or in combination

The main and obvious advantage of secondary heat sources is heating the space in which they are located. These were not only just effective however in providing space heating. However, we found three other significant benefits from our heat pumps combined with multi-fuel stoves in addition to providing localised space heating:

1. Reducing the overall central heating operating time.
2. Heating rooms inaccessible to the main central heating system.
3. Back-up function.

1. Reducing the central heating operating time

The major benefits of different combinations of heat pumps and multi-fuel stoves in terms of central heating operating time are probably best illustrated by examples of how we use these in the Farmhouse and Barn Conversion.

- Farmhouse case study: the use of renewables and secondary heat sources combined

We have no large spaces to heat in the Farmhouse and heat pumps therefore only had a marginal benefit in terms of space heating alone. As these primarily heat the area in which they are located with modest convection to other rooms, they are of limited use in a small individual room unless this is heavily used. We have therefore made more use of multi-fuel stoves. We installed one with a back boiler heating

145

three radiators in the core of the Farmhouse at the time of the original build, providing a heat output of around 3kW of direct heat to the main room and up to around 10kW from the back boiler through the radiators. This was followed by a 4.5kW output air-source heat pump, installed in the heavily used Farmhouse Extension, because it both provided inexpensive space heating with low carbon emissions (albeit to a small area) and because the position of the extension made it difficult to easily access the central heating system to plumb in a further radiator. We added another multi-fuel stove without a back boiler in the kitchen area to replace an inefficient wick-type oil-fired Rayburn. This stove provides around 6kW of direct heat.

Generally there are two of us in the house and in cold weather the central heating system would be run throughout the day and evening. We would typically use and require heating in three main areas of the Farmhouse:

→ The kitchen, upper mezzanine office and Farmhouse Extension set of rooms – during daytime hours;

→ The living room and main bedroom immediately above it – in the evening; and

→ The main bedroom – first thing in the morning and late evening.

As our children now live away from home we set the TRVs to minimum in the other bedrooms in the house when they are not at home. Apart from the central heating system, the kitchen/ mezzanine/ extension set of rooms can be fully heated by the 4.5kW output air-source heat pump and the 6kW multi-fuel stove (without a back boiler). Again, apart from the central heating system, the living room/main bedroom set of rooms is heated by the multi-fuel stove and back boiler, which is plumbed into three radiators – one in the sitting room itself, one in the main bedroom and one in a second bedroom which is used only when one of our children is at home and is turned off at other times (around 13kW of potential heat in total).

Each room set therefore has secondary heat sources providing upwards of 10kW, ample to heat the areas concerned. The overall heat pump contribution is limited to the one room, the Farmhouse Extension, although this space is one of the most heavily used rooms during the daytime. The heat pump and stove also provide additional heat to the neighbouring kitchen and mezzanine office/extension and the Farmhouse Extension respectively.

This combination of heat sources together with replacing the Rayburn reduced our oil use from an annual cost of approximately £4,000 to £1,000 before the start of the war in Ukraine. The alternative fuel sources consumed a total of around £700. All of these costs have risen, oil far more than the alternatives (wood is almost unchanged), and the savings are now therefore far greater.

- Barn Conversion case study: the use of renewables and secondary heat sources

This build was suitable for a central heating zone valve to isolate the two parts of the building, the gin gan and the large kitchen/dining/sitting area and bedrooms/bathroom. We felt that trying to introduce a second valve to separate the kitchen/dining areas from the bedrooms/bathroom would overcomplicate the system.

During an average day, and not unlike the Farmhouse, different areas of the Barn Conversion are used:

→ The kitchen/dining and upper mezzanine area set of rooms – mostly but not only during daytime hours and first thing in the morning;

→ The gin gan area – in the evening; and

→ The bedrooms/bathroom area – early in the morning and late in the evening.

Apart from the central heating, the kitchen/dining/living area is heated by the 7.5kW output air-source heat pump. Again, apart from the central heating, the gin gan is heated by the 4.5kW heat pump and 15kW output multi-fuel stove. The bedroom area had no secondary heat source.

Each room set, kitchen/dining area and the gin gan, therefore has sufficient secondary heat sources providing 7.5 and 14.5kW of heat respectively, enough to heat most of the two areas. Some heat is provided to the lower bedrooms by convection from the kitchen/dining area heat pump, although the pump provides sufficient heat to the upper mezzanine and bedroom from the greater effect of heat rising by convection than in the lower bedroom area.

This arrangement has less impact on reducing the central heating 'on' time than in the Farmhouse as the large heat pump does not provide all the heat required for the kitchen/dining area in cold weather and the heat provided to the lower bedroom area is limited. It does still greatly reduce the central heating operating time and the oil usage.

Both the heat pumps and the stoves provide heat rapidly, the pump somewhat faster than the central heating system and the stoves somewhat slower.

2. Heating rooms inaccessible to the main central heating system

This has been detailed for the Farmhouse Extension (see Page 125) and was the solution we also chose for a converted outbuilding (which I have not described). In each case, the only realistic alternative would have been an electric radiator system, which would have been far more expensive to run (although cheaper to install) and have a far larger carbon footprint. As the systems are simple to install (taking around a day) they suit this situation extremely well.

3. Back-up function

In both the examples described above, the buildings are heated by three sources: the central heating system, the multi-fuel stoves (one with a back boiler in the Farmhouse) and the heat pumps. Most power source failures can, therefore, be backed up to variable degrees.

- Boiler failure (or other failure in the central heating system) can be compensated in the two main areas of the Farmhouse and in the Barn by the multi-fuel stoves and/or heat pump(s).
- Heat pump failure (rare in our experience except at the end of life) is compensated by either the central heating system or multi-fuel stoves in the Farmhouse and Barn.
- Electricity outage is more problematic: this will obviously paralyse both the heat pump and central heating system, but also any multi-fuel stove with a back boiler, which needs an electric pump to distribute water to the radiators. In the Farmhouse, the multi-fuel stove without a back boiler in the kitchen room set will provide (just) sufficient heat for the kitchen, mezzanine and Farmhouse Extension. We have found that we can still also use the stove with a back boiler in the sitting room very cautiously (to avoid overheating and damage to the system) in order to provide some background heat to this room. In the Barn Conversion, the multi-fuel stove in the gin gan will heat this room and provide some additional heat to the kitchen/dining area.

Net effect

The combination of these additional heat sources therefore provides ample heat, minimises our use of the central heating system (and consequential carbon footprint), heats areas inaccessible to the central heating system (the heat pump) and provides back-up functions. Both the stove and heat pump provide rapid heat and can be turned off (or allowed to run down) at will, greatly reducing or avoiding the need to use the central heating system.

The arrangements described in the Farmhouse and Barn Conversion are only examples based on how these properties are used, and many combinations and permutations can be installed to achieve a similar effect. We have the potential to generate around 25kW in each of the Farmhouse and Barn Conversion, which is therefore almost an order of magnitude greater than the average amount of energy which can be used from wind or solar power and is far cheaper to install, in addition to the other benefits described above. Taking account of the net energy generation from the heat pumps of around four times the energy consumed to run the systems, and the fact that unless used otherwise, wood left to decompose would generate the same amount of CO_2 as burning it, without the heat, the 25kW generated is far better than carbon neutral.

It is difficult to quantify the precise payback time of the heat pumps and secondary heat sources, although in our case this was no more than a few years in view of the reduction in our oil consumption. Although the heat generated from the heat pumps and stoves is significantly cheaper than using oil (or gas), a large part of the cost-benefit is due to the reduction in the central heating 'on' time and avoiding heating rooms unnecessarily rather than space heating alone. These units are not particularly visually intrusive and operate quietly, the main source of noise (although limited) coming from the outside fan unit (**Photos 124 and 125**).

Maintenance and running costs: impact of build solutions

We have had remarkably few maintenance issues, which may be partly because of some of the decisions we took. In particular, we have avoided repeated pointing of walls over the years and not had significant roofing issues; both potentially high maintenance costs. We have had to fill or repair occasional patches of flooring

Photo 124 Air-source heat pump (air-to-air). External fan unit.

Photo 125 Air-source heat pump (air-to-air). Internal unit.

which we had cleaned and restored rather than replaced, but we had expected this and filling a few holes is not a difficult process. We have not had any significant deterioration since. The only other maintenance any of the buildings have needed has been routine work such as repainting or recoating external wood (windows, doors etc.).

Modern versus old approaches to the buildings

Ignoring the Garage, which was a modern building, we used a combination of approaches depending on the building space. In two, which contained a wealth of original features, we chose to keep the 'old' appearance completely (the Farmhouse and Barn). In others, we opted to create a more modern appearance internally (Mill, Byre, Farmhouse Extension and Barn Conversion) but retained the main original features. Changes to the exterior of the buildings were kept to a minimum and in most cases any additions proposed to the building (a morning room in the Mill and the Farmhouse Extension) involved rebuilding structures (however rudimentary) which had been present previously, or creating openings where previous ones had existed. For planning purposes it was useful for these to have evidence, either from existing remaining parts of the addition (such as the dilapidated entrance structure in the Mill and an old aerial photograph in the Farmhouse) or from historical changes (such as the lintels and closed wall sections in the Barn). We also preserved the heritage of the Georgian and far older parts of the Farmhouse and the Victorian appearance of the adjoining cottage, making almost no changes to the internal fabric of any of the buildings but leaving these in keeping with the different eras.

We found the conservation planning officer, the English Heritage officer in the case of the Farmhouse, and the planning authorities in general, to be extremely helpful and had few difficulties. We were able to resolve any minor details by gentle negotiation, helped I think by the fact that we had worked closely with them and been completely transparent about our wishes. Whilst we may just have been lucky, these positive experiences have covered many projects at different times, working with different planning and building control officers over a period of 20 years.

How a building is converted or restored, whether 'old' or 'new' is a matter of personal preference. It is sad, however, to see period buildings 'modernised' to bad effect. During our house hunting we did come across a number of restorations and conversions where the owner had used expensive modern fixtures and fittings throughout an old barn or house. Some modern mahogany skirting boards, despite being expensive, did not sit well at all. At its worst, we came across a barn conversion in which a hugely expensive but equally garish bathroom suite had been installed, more suited to a banana republic palace than an old building!

I would finish this chapter therefore with a personal plea: if you are intending to create a new finish in an old building please cover up the bare minimum of original features and think carefully about how the new will sit alongside the old before plunging into any one decision.

Chapter 6: Results in practice

This chapter examines results, durability, costs and the savings we were able to achieve at no detriment (and in many cases arguably an improvement) to the final result.

Were the intended results achieved?
Immediate results

In the Garage, we wanted to create a simple one-bedroom dwelling out of a former new garage. We had sufficient space to create an open plan kitchen and living room area together with a small entrance hall – part of which was given up to a spiral staircase in order to access an upstairs bedroom as we had sufficient height to create an upper floor room. Given the layout, it would have been difficult to use a conventional staircase and although spiral staircases take up more room than might be imagined, in this case ours proved extremely useful. As the building was modern, it had no period features. We obtained permission to add sufficient windows on the side of the building (not overlooking any other property) to provide sufficient light and the final external and internal results were exactly as intended.

In the Mill, our intention was to create a mostly modern internal appearance consisting of an open plan lower floor with bedrooms and bathrooms on the upper floor and a modest extension offering an additional warm living space at the entrance to the upper floor. We kept the old mill workings at one end of the building and a former large fuel hopper at the other. The kitchen created in the lower smaller area provided ample space to include a breakfasting area by cutting the kiln but preserving at least the appearance of the former inside of the structure. The partitioning on the upper floor to create the bedrooms and bathroom/toilet was necessary and achieved as planned and the open staircase between the floors provided a clear view through the tensioned wiring towards the mill workings. Apart from the external pointing and the structure of the extension, we achieved a final result which was very close to what we had intended. The pointing and extension issues were both due to failings on our part as we did not appoint a project manager to make regular checks on the work.

In the Farmhouse, our intention was to keep the inside of the building as close as possible to the original, but opening the upper ceilings by removing the badly damaged plasterboard which had been added in the 20th century to re-expose the roof timbers once these had been discovered. We also wanted to rearrange the long extension, probably added in the Victorian era, to create a more functional bathroom, a more usable space in the main part of the extension with a laundry/utility room and a further bathroom at the end of the space. This did not involve any change to the structure of the building. Creating an opening from the Farmhouse into the smaller area of the Barn provided additional (kitchen) space. This building was unusual as we discovered medieval features which were unknown at the start of the build. We were able to preserve these, although a bit of rapid thinking was required. The final result, as far as we were concerned, was excellent. The main change to the inside of the Farmhouse was to expose the medieval findings and roofing timbers in addition to preserving the appearance of the original Farmhouse itself. We were able to make these changes successfully, much as planned originally apart from work on the medieval features which we had to accommodate along the way.

In the original Barn, we had intended to leave the space completely open but to divide the large space without walls or partitions into a small snug beneath a new mezzanine, a dining area and a larger sitting area. Similarly, we left the gin gan open. The roof timbers were left exposed throughout the Barn and its gin gan and the only additional structural change was to add three rooflights and create a mezzanine at one end of the Barn, mirroring the mezzanine above the smaller end (now the Farmhouse kitchen). Again, like the Mill, the tensioned wiring on the staircase in the larger part of the Barn kept the appearance as open as possible. This was therefore a simple build and we achieved what we had set out to do.

In the Byre, we had intended to create a dwelling similar to the Garage but with the bedroom on the lower floor as we did not have sufficient height to create any accommodation above because of the roof level and structural A-frames. As we left some of the animal stall concrete dividers in place and exposed an external wall, we were able to keep a number of the features of the original byre. The final result was again very much as planned.

The Farmhouse Extension was a simple build, finished as planned, and was true to the original appearance of the building from an earlier aerial photograph. In particular, we used corrugated sheets for the roof in order to match this with the original appearance of the add-on. The only problem we had was heating the room, as it was difficult to access from the main Farmhouse central heating system. However, we had no difficulty in solving this problem using an air-source heat pump, which also reduced our running costs significantly. This room has become one of the most heavily used areas in the house and is attractive and warm (as it is south-facing). It has been highly successful and, in our opinion, a very worthwhile and inexpensive investment.

The aim in **the later Barn Conversion** was to create three bedrooms and a bathroom in the original Barn space but to leave the building as open as possible in order to retain the overall appearance of a barn, although with a modern interior. To do this we intended to use approximately half of the lower Barn area to create the bedrooms and bathroom, keeping the original staircase up to the mezzanine to access a further bedroom. We succeeded in leaving the overall appearance open, this time using toughened glass panels for the additional staircase and balustrade at the end of a new mezzanine overlooking the rest (approximately half) of the Barn above the lower area. We combined this with a large glazed panel in the bulkhead between the new mezzanine and the upper bedroom, again to leave the building as open as possible. The area leading from the Barn to the Byre created a small but useful utility space housing a new boiler and white goods. Our final result was exactly as planned.

Durability

We carried out the first major project (the Mill) in 1994 and we sold the building in 2014 in much the same state as it was in when we had completed the conversion. Apart from routine maintenance, no problems arose during the 20 years we owned it. We carried out the second major set of projects between 2000 and 2002, and again, apart from routine maintenance and a few problems which were of our own making (notably plastic plumbing pipes), the property as it stands in 2022 has stood the test of time for over 20 years. We carried out the later projects between 2007 and 2016 and we have not yet needed any further work other than basic routine maintenance.

Old Buildings: Conversion and Restoration

How much did they all cost?

This is a difficult question to answer to the penny. We were perhaps a little lax in keeping an exhaustive record of every cost, so it is difficult to estimate the exact spending from invoice records.

Taken from multiple estimates readily available on the internet, the lowest average cost quoted for new builds outside of London at 2020 prices is approximately £1,200/m^2, compared to a figure of £1,500/m^2 for converting a barn.

Again from published estimates, the cost of converting or restoring a Listed building can be anywhere up to 50% greater.

It is impossible to find a true benchmark against which to compare our experiences. Because of the wide range in reported estimates, the average figures quoted only give an overall guide. In terms of avoided costs, it is probably most realistic to add together all the individual savings we made and compare the total to what we would have spent had we gone down the purely conventional route – doing all of the works suggested by an architect and not taking the decisions or finding the materials we did. In each of the projects the final cost was well below half the lower quoted estimates before taking into account that, apart from the Garage, all the buildings were Listed.

The **Garage** project cost approximately £20,000 at today's prices: for a final floor area of approximately 50m^2, ≈ £400/m^2. The building however was new and insulated and had a damp proof course.

The **Mill** cost approximately £140,000 at today's prices: for a final floor area of approximately 210m^2, ≈ £660/m^2. This was totally derelict originally with little or no services, although the building itself and most of the roof were sound.

The **Farmhouse** restoration cost approximately £170,000 at today's prices: for a final floor area of approximately 260m^2, ≈ £650/m^2. This was originally also totally derelict but with services on site and both the buildings and roofs were in good condition.

The **original conversion of the Barn** at the time of the initial farm complex projects cost approximately £70,000 at today's prices: for a final floor area of approximately 160m^2, ≈ £440/m^2, with the building and roof in good condition. Apart from the mezzanine, however, this was originally left as a large open space without separate rooms, kitchen or bathroom and this cost does not, therefore, reflect the cost of a barn conversion into a usable dwelling.

Results in practice

The Byre conversion cost a total of approximately £30,000 at today's prices: for a final floor area of approximately 70m², ≈ £430/m². Like the other farm complex buildings this was totally derelict with a poor roof but sound walls.

The Farmhouse Extension cost approximately £12,000 at today's prices: for a final floor area of approximately 20m², ≈ £600/m², for a simple new build.

The Barn Conversion cost a total of approximately £50,000 at today's prices: for a final floor area of approximately 190m², ≈ £260/m², although much of the work required for the building had been done at the time of the original Barn project. In terms of a full barn conversion with different rooms, kitchen and bathroom(s) carried out from scratch, adding the Barn and Barn Conversion costs together would give a total overall cost of ≈ £610/m², with all services on site and both the building and roof in good condition.

As can be seen from the above figures, the Farmhouse itself was, as expected, the most expensive restoration because of the need to strip out all the existing wall coverings and lay screeds over a damp course membrane throughout the lower floor rooms, raise a lintel and restore all the existing six rooms, halls and the Victorian extension.

The reasons behind many of our decisions were not based on costs alone but were aimed to improve the effect on the final result. These decisions did also produce large savings compared to their new or specialist company equivalents.

The renewables and secondary heating options we chose were inexpensive, particularly in terms of the heat they produced. The heat pumps cost around £1,200–£1,500/pump in both original and today's costs, as these have not changed much over 15 years (in our own experience). The same applies to the multi-fuel stoves, which

Photo 126 Another major overdraft contributor – three daughters who love musicals: farm conversion versus seats to see Hamilton!

cost £1,500–£2,000 each for a stove with back boiler and radiator(s) installed and £1,000 for a stove alone. Although we installed these at different times, in the initial build and several years later, the total spend amounted to around 2% of the total build cost for the Farmhouse (one heat pump and two stoves) and 5% in each of the Barn (one heat pump and one stove with back boiler) and the Barn conversion (an additional heat pump and relocating the original Barn stove).

Where were the savings made?

The savings made derived mostly from the principles and ideas listed in Chapter 2 and described as they applied to the different project types in Chapter 4. Some of these applied to all of the projects, although each principle or idea contributed in different ways and made different savings in each project.

Two areas in which large savings were made in all of the projects were our different approaches to instructing an architect or using an alternative service (planning consultant or architectural draughtsman) and using individual tradespeople.

Our use of architect, planning consultant or draughtsman services varied between projects and are described for the individual buildings.

We mostly used individual tradespeople in the different projects to:

1. reduce VAT (to a variable extent),
2. avoid the higher costs of using a larger contractor (loosely estimated as 25–30%), and
3. be able to work with our tradespeople directly.

The savings made varied between projects mostly because of the VAT status of the building. Together these two types of savings reduced overall costs in our projects by somewhere between 30–50% before any of the other savings are taken into account.

In order to use these approaches an owner must feel confident both in building design and in project management, although advice from a specialist can be built in without handing over the whole process to someone else. In our experience this was the most rewarding part of each project and good or bad (I hope good in our case), and despite some difficulties in the project managed from a distance, we would not hesitate to do things the same way again. There is an enormous amount to be learned from this and some lifelong friends to be made at the same time

I have made a number of assumptions to calculate savings, some of which I have described in Chapter 1, Assumptions; these are summarised below. Some, such as the additional cost of using larger contractors compared to individual tradespeople are somewhat arbitrary but I have based these on differences in quotations where we had these to compare. As our own labour costs for individual tradespeople (£15–£20/hour) are close to or over the UK average wage, a figure of 25% is likely to be conservative once premises, salaries of back-office staff and directors, advertising and profits and/or dividends to shareholders are taken into account for larger contractors. Other assumptions such as VAT liability are fairly accurate. Some costs (such as materials and using a local fabricator) varied between projects depending on the amounts (e.g. flooring) or types (e.g. Georgian-style sash windows) required. I have not included any savings made from our choices of kitchen and bathroom fittings as these are a matter of personal taste. In our experience, however, 'inexpensive and simple' produced a perfectly satisfactory result which did not intrude visually on the conversion or restoration; although if fixtures and fitting are purchased from specialist suppliers this can easily add a further 5% to the total build costs. The costs of adding renewables or secondary heating sources are limited and not included, on the basis that these are investments made to reduce running costs.

Assumptions and figures used to calculate savings:
- VAT liability is 5% for a VAT-registered contractor for qualifying works and 20% for non-qualifying works.
- Approximately 30% of potential costs are for materials.
- A larger contractor working from a quotation costs approximately 25% more than individual tradespeople working on a day rate basis.
- The current cost of using an architect to design, create plans, draw up specifications, put these out to tender and oversee the works is approximately 10% for an unlisted building or around twice that for a Listed building; this is approximately three to five times more than using an expert (architect or other) to draw up plans only, with the specifications being produced (with the help of a trusted builder or tradesperson) and the project managed by the building owner.
- The costs of avoided works or using alternative building techniques are based either on actual quotations received where these existed or from estimated averages found from internet searches.

- The savings made from reclaimed materials (or repairing rather than replacing) are based on actual costs where these are available and compared to quoted averages either from a local supplier or from an internet search.
- The costs of a local fabricator to produce materials or items (such as windows) are approximately 50% less than using a specialist company.

In the Garage project, we achieved our main savings by:
1. Using an architectural draughtsman and our own design, saving around 10%.
2. Using individual tradespeople compared to a larger contractor, saving in the region of 25%.
3. Designing, drawing up specifications and project managing ourselves, leaving the plans themselves to an architectural draughtsman, saving around a further 15% compared to the cost of instructing an architect for the whole project.
4. Using individual tradespeople also limited VAT liability, saving around 15% in today's money as VAT was still payable on materials.
5. Compared to a large contractor, using individual tradespeople reduced costs by a further amount in the region of 25%.
6. Making modest additional savings using reclaimed materials, although these still amounted to around 10% of the total potential build cost.

In the Mill project, we used an architect to produce the plans and initial specifications. We achieved our subsequent main savings by:
1. Project managing ourselves, or trying to, saving around 10% of total potential build costs. In retrospect, this was a mistake because of living a distance away and it is perhaps unrealistic to include this as a saving.
2. Avoiding some works originally proposed by the architect, but thought not to be strictly necessary by our surveyor, saving around 15% of the total potential build cost.
3. Reducing the total potential build costs compared to a larger contractor, by using individual tradespeople, saving somewhere in the region of 25%.
4. Using individual tradespeople to limit VAT liability, saving around 5% compared to the VAT liability for a larger contractor but no saving for a self-build under 2022 VAT rules.
5. Using reclaimed flooring and some wood for finishing joinery, saving in the region of 5–10% of total build cost, although few other reclaimed materials were used.
6. Using a local fabricator for joinery parts, saving in the region of 10% of total build costs.

We used similar approaches in the subsequent projects, notably working with individual tradespeople, although the contributions made by the different types of savings varied between projects, depending on the works described in the case studies in Chapter 4. The main differences between the savings in the different projects are summarised below.

1. We only instructed an architect to draw up and submit plans for all of the later projects, apart from the Barn Conversion when we used a planning consultant to assist with design and to draw up and submit plans. In each case, this saved around 15% of total potential costs.
2. The savings from avoidable works varied considerably, from around 20% of build cost in the Farmhouse and Barn to none in the Byre, Farmhouse Extension (a new build), and Barn Conversion (essentially a continuation of the Barn project to create a separate dwelling, carried out many years later).
3. Using alternative building techniques, notably how we treated the interior and exterior of the external rubble-filled walls in the three projects on the original farm complex (Farmhouse, Barn and Byre), made very considerable savings, in the region of 10% of total potential build costs. We did however also make a significant saving in the region of 10% of total cost by reproducing the roof style (corrugated sheeting) rather than tiling in the Farmhouse Extension.
4. As in the previous projects, the cost of individual tradespeople was probably in the region of 25% compared to a large contractor.
5. The VAT savings ranged from none in the Farmhouse, Barn and Byre as these were self-builds and not therefore liable for VAT under 2022 rules to just around 15% in the Farmhouse Extension and Barn Conversion, (VAT was payable on materials but not on the works by the individual tradespeople). The VAT benefits of using non-VAT registered tradespeople are therefore entirely project-dependent.
6. The greatest savings from using reclaimed materials were, not surprisingly, made in the two projects on the original farm complex (Farmhouse and Barn) where we were aiming to preserve the old appearance of the interior of the buildings (around 15% of total build costs). In the Byre and the two later projects (Farmhouse Extension and Barn Conversion), we designed modern-style interior finishes and the savings were more modest (around 5%, mostly from reclaimed flooring in the Farmhouse Extension).
7. Although we did not intend to use 'secondhand' modern items such as boilers or oil tanks, because of the risk

they carry, we did find these from a trusted source, adding a saving of around 10% of build costs in the Byre and Barn Conversion projects.

8. Finally, and not surprisingly, the greatest savings from using a local joinery fabricator were made in the Farmhouse (around 5% of total potential costs), as this was the project which required the most complicated woodwork, notably windows and machined wood for doors and other finishing joinery.

Our top and bottom 10 decisions

For the most part our decisions have worked and produced a better final appearance than we might have hoped for, and at a considerably lower cost than expected. Some decisions were far better than others and a few went spectacularly wrong! Below is a list of the top and bottom 10 things which we either did or did not do, together with a brief discussion of each. The top 10 had a huge impact on both the final result and on build costs, whereas the bottom 10 were mostly minor factors and had a far more limited effect. This is probably because the 'top 10' together had such a positive effect on the builds that the 'bottom 10' only reflected minor mistakes or omissions.

Top 10

Several of these may appear obvious and some contributed far more than others towards reducing our build (and also running) costs.

1. Finding the right builders and using individual tradespeople

This one is hardly surprising and was responsible for most of the savings made and the quality of the final results. We were successful with this in the Garage, Farmhouse, Barn, Byre, Farmhouse Extension, and Barn Conversion, where we were very fortunate in knowing reliable builders. We were less so in the Mill. The concept of six degrees of separation linking everyone on the planet has clearly worked, at least for us with 'The Boys'. Whilst our farm projects were the first they had taken on since setting out to work independently, a large number of friends and acquaintances have worked with them since. This confirms our own feedback on the works done for others.

2. Reviewing our use of architects and other design specialists

We were aiming to retain the appearance of the buildings, or in the case of the outbuildings (Barn and Byre) to leave the appearance as true to the original as possible, rather than creating an externally striking 'statement' design. As our designs were all therefore straightforward and we felt confident to project manage ourselves (apart from the Mill), we used the services of an architect only to draw up plans from our designs and submit these to the planning authorities, and to provide one-off guidance when we felt this to be necessary. In one case (the Barn Conversion) we did this through a planning consultant and included aspects of the design in this case. We found these approaches to be entirely satisfactory and reduced our overall costs by around 15% of total build costs.

3. Paying by the hour

This is more controversial but proved to be effective for us. Although it carries some risk, if there is a bond of trust between owner and builder this process avoids the need to work to specifications defined in advance and therefore enables the tradespeople to respond immediately to problems encountered along the way. It allowed us to deal with unforeseen circumstances directly with them rather than having to ask for further quotations during the works which we would have had to do if they had been instructed on a quctation basis. We cannot estimate any savings (or additional cost) of using this approach but it certainly allowed flexibility. All that I can say about costs, is that from daily supervision of the works they appeared to progress remarkably quickly.

4. Identifying and avoiding major works which were not strictly necessary

We looked at the works required extremely closely with both our builders and experts, but did on occasion look for a second opinion on some works, such as the (expensive) recommendation to tank a wall in the Mill and the extremely expensive option of rebuilding the medieval wall in the Farmhouse to incorporate supporting stainless steel rods. We did not look for an expert opinion about reroofing the farm complex of buildings but went along with our own initial impression and our roofer's own advice that only one smaller section of the complex (the Byre project) called for the building to be reroofed. This reduced costs enormously and has not caused us any problems over more than 20 years.

5. Relentless quest for materials

Extending our search for materials far beyond the obvious (wood flooring) led to some major finds in terms of cost and appearance. (insulation, wooden guttering, boilers and oil tanks, multi-fuel stoves, an array of wood for structural timbers and finishing joinery, Caithness flagstones, external doors, kitchen units, a spiral staircase and furniture – the latter more of a very useful by-product of the search for materials). These had five figure impacts on cost and, arguably, a positive effect on the finished appearance compared to buying new (hours spent researching possible materials paid back many times over). Both eBay and Facebook Marketplace (beginning with local searches and then extending these) as well as local papers are good starting points. The advent of internet sources has also brought with it many cost-effective means of transporting even large objects.

6. Carefully considering building techniques and options

This also had a major impact. We greatly reduced, for example, both the materials and labour costs of pointing the exterior of the building. As described earlier (see Page 90) the concept of a wall 'breathing' fully does not necessarily apply to a wall with a cavity filled with small amounts of rubble. In our buildings we found that a 'weak' pointing mix, made on site, containing lime but also a small amount of cement rather than a proprietary pre-made lime putty mortar gave a sound and attractive result. This has lasted well over 20 years with no detrimental effect in terms of damp or building fabric. Similarly, constructing a roof from corrugated sheeting in the Farmhouse Extension to match the earlier appearance greatly reduced costs but also improved the visual result. As a third example, portable sandblasting in the Farmhouse had a large effect on the finished appearance and allowed us to avoid expensive and potentially less sympathetic replacements. It was helpful both for its conventional use on metal and stone but also helped enormously when used cautiously on wood, allowing us to retain original structures (staircase and flooring) which would have otherwise been unviable.

7. Careful examination of renewables and other secondary heat sources

We could easily have spent well over £50,000 on wind and solar electricity generation and ground, air-source or biomass heating combined on any one main project. For us, the most financially-viable and environmentally-pragmatic compromise was air-to-air heat pumps and woodburners with and without a back boiler. The latter tackle the important issue of small particle emissions, although these have been hugely reduced by installing a new

Ecodesign stove and only burning 'dry' wood. There is an obvious trade-off to be made between burning wood and fossil fuels. In the Farmhouse and Barn, for example, this combination produced a potential of 4.5kW and 12kW of heat generation from the heat pumps and 20kW and 15kW from the multi-fuel stoves, with or without a back boiler system, respectively. This dwarfs the potential of wind and solar generation even under optimal weather conditions (only rarely achieved), at 20% of the cost. This combination has reduced our oil consumption in the Farmhouse for example by two-thirds, or 12.5 tonnes of CO_2 annually, compared to the time when we were not using these sources systematically in a strategy to reduce our consumption of oil. Owners should, however, check the incentives available at the time.

8. Making our own woodwork either on site or with a joinery fabricator

From our own experiences, windows, staircases, doors, door jambs and architraves and, in the case of the Mill, even a customised cupola made by a joinery fabricator, enabled us to remain sympathetic to the original appearance of the buildings at a far lower cost compared to using specialist manufactures. These people can work wonders with various types of reclaimed wood.

9. Salvaging existing materials and never throwing anything out!

Even if slightly ragged at the edges, old salvaged materials (such as doors and door furniture suitably cleaned and renovated where necessary) both greatly reduced cost and helped to recreate the original appearance of the buildings. Do not throw anything out! A use will always be found for leftovers. It is also easy to forget the cumulative cost of sending people off to buy assorted additional small materials, both in terms of the cost of the individual materials themselves and of the tradesperson's time.

10. A bit of thought regarding disposal of rubbish

This may seem a minor consideration, but disposal of rubble is an expensive process. Our buildings, particularly the Farmhouse and Barn, generated enormous amounts of rubble, which together could have potentially cost up to five figures to remove. Using this for obvious purposes such as hard-core for drives, hard standing areas and paths, but also burying large amounts to level or raise garden areas, was an inexpensive process involving a day or so of a JCB.

Bottom 10

Fortunately, the majority of our decisions have proved, both at the time and in retrospect 20 years later, broadly successful. Several of the bottom ten are slightly peripheral to the main build but are worth highlighting anyway.

1. Not appointing a project manager for a remote build

Visits by a project manager once or twice weekly to the Mill would have allowed us to identify some of the difficulties faced by our first builder before his problems had become more apparent. This was our greatest mistake and was a naïve oversight, although our options would have been somewhat limited at the time by the relative remoteness of the building. In retrospect, instructing a project manager or architect would probably have been the best choice at the time.

2. Overstretching our tradesperson

The main source of difficulty was in the project that was managed remotely (the Mill). Our experiences in the farm complex set of projects with our tradespeople were overwhelmingly successful as the individuals concerned had all of the trades skills necessary. Our failing in the Mill was that we instructed our 'general builder' to proceed from the masonry works, which in the most part he carried out satisfactorily, to the first fix joinery which was beyond his competence. We could perhaps have picked this up by having a more detailed discussion with him in advance of asking him to carry out the work or by arranging regular site visits (as in 1 above). This was a risk we had taken in appointing individual tradespeople rather than a contractor with dedicated stonemasons, joiners and electricians etc., and in this case, the gamble proved unsuccessful. There are no fail-safe means of avoiding this possibility, other than choosing a larger contractor offering a range of skilled tradespeople, but this would be at additional cost; given adequate knowledge an owner should be able to reduce the risk by having a careful discussion of the specifications with the tradesperson to gain some insight to their own knowledge and skills.

3. Poor choices in plumbing and electric materials

Plastic plumbing materials were incompatible with our building because of damage from rodents which resulted in major problems with leaks. Unprotected electrical wiring was probably also a mistake. Using conventional copper plumbing materials and protecting electrical wiring with a first line of defence (conduits or armoured cabling for vulnerable electrical wiring) would have minimised the potential for damage. Similarly, an early decision in the Barn to embed electrical sockets in the floor surface to reduce the visual effect of mounting these on walls was unwise, as it left the sockets exposed to minor dust fall from walls or in some cases from a cat relieving itself!

4. Not separating services between buildings

At the time of the original build we had not intended to divide the Barn off from the Farmhouse or to register the Byre as a separate property. We did this some 20 years later. Truly separate properties need separate services. The necessary cabling and pipework is easy to install during the initial build when service trenches are open and before flooring has been laid, but expensive and disruptive to do at a later stage. Where a building or set of buildings contains interconnecting parts, which could potentially be separated later, considerable cost and disruption can be avoided by installing separate services at the time of the original build.

5. Not giving enough thought about light

In the Mill, we had no particular issues with light apart from the lower floor where the window openings were north-facing and the main room area was potentially rather dark. A small void to light part of the upper floor easily resolved this. In the Barn, however, the main openings were at one end of the building and we failed to give enough thought to the other and only installed a single rooflight, which offered limited extra light. We did have two areas of closed up external wall at the other end of the building which we later opened at the time of the Barn conversion; opening one of these at the time of the original build would have avoided this problem at minimal cost.

6. Failure to consider flooding

We had assumed that we did not have the potential for flooding in any of the properties, as none of the main buildings was at risk from rising river levels. However, we had problems on both of the main sites, the Mill and the farm buildings, as we had not considered the impact of run-off from distant hills directly into the foundations of the buildings. A couple of simple grate drains and modest works to divert possible run-off resolved these problems easily at a later stage. In hindsight, we should have considered this at the time and taken action at the build stage.

7. Buying an old-style Aga/Rayburn stove

Despite looking good, an old-style Aga-type stove (with a wick burner) consumes an enormous amount of oil. We should have done far more research. Modern versions with a condensing boiler are infinitely more efficient and could be considered if the money is available. An owner may however prefer to have a dedicated condensing boiler, in which case an Aga-style range is an expensive indulgence for cooking alone and just providing heat to the room (kitchen) in which it is installed. The older stoves using a wick were extremely expensive to run and, in our experience, not particularly effective or responsive. If an owner wants a range as a centrepiece to a kitchen purely for cooking it would probably be far less expensive and more practical to opt for one powered by electricity. Although more expensive to run than mains gas or oil, the absolute cost of heat for cooking is modest.

8. Being slow in taking up renewables

I think our choice of renewables was the most practical and cost-effective at the time we installed our systems. We were perhaps amongst the earlier people to adopt air-to-air heat pumps (having dismissed the other renewables as not being economically viable), although we could and should have researched these even earlier to maximise the benefit we obtained from them. Likewise, earlier research into solar electricity generation could have allowed us to lock into the generous feed-in tariffs available years ago which would have made installing solar panels a far more financially attractive option if we had managed to negotiate planning permission. It would be wise to keep abreast of options for renewables as these develop and factor in the government incentives schemes as they apply at the time of a build. Bear in mind, however, that estimates of the potential benefits of solar and wind systems usually do not take into account the lifespan of the equipment, which is not limitless, and in view of the initial capital outlay, depreciation costs will reduce or may even negate the value of the electricity generated for personal use.

9. Not installing sufficient electricity switches and sockets

This probably applies to all building projects and in our case particularly applied to the Farmhouse with its predefined rooms. The adage that one can never have too many switches and sockets is quite true. We had not given enough thought to where electrical items would find their resting place or to the direction of footfall throughout the building, which cannot always be reliably predicted in advance, and was not in our case. Several more electrical sockets and additional light switches would have been beneficial and inexpensive to install at the original

Results in practice

build stage; in practice, this is almost impossible later on without considerable disruption.

10. Using old-style 'breathable' casein-based paints when this was not necessary

We used casein-based paints on the internal surfaces of exterior walls in the Farmhouse in the belief that the colours would be more traditional and that this would increase the ability of the walls to breathe (both incorrect assumptions).

Any water-impermeable layer, including modern emulsion patents, reduces or abolishes the ability of a lime mortar wall to fully breathe and the wick effect of lime to allow water vapour to pass through the wall and evaporate. If the wall construction does not absolutely need to be finished with a lime mortar and pointing (in our case because of the rubble-filled rather than solid walls), the same finishing colours can be achieved equally well with commercial 'non-breathing' emulsions.

As our reconstituted casein-type powder paint was intolerant of cold, it went 'off' and we found it impossible to create a colour-matched mix to touch-up paintwork.

Repainting a room also became a far more complex and expensive exercise involving dusty rubbing down of the paintwork to a completely smooth finish and the application of two coats of stabiliser to all of the walls before repainting.

Chapter 7: Footnotes: things that happened along the way

This chapter is included more for interest and fun rather than guidance and is essentially a group of outtakes from all our building projects from the early group of flats in Glasgow in the 1980s to completion of the Barn Conversion, in the year 2017. It recounts a few experiences over that time, some of which are instructive and others are purely of interest or fun. Everyone undertaking any form of restoration or conversion will encounter gremlins along the way: these present their own challenges but in many ways enrich the whole experience provided one has humility, a sense of humour and self-deprecation.

The Barras

Some of my earlier forays into property renovations included properties in Glasgow. One interesting project in the west end of Glasgow, carried out while I was living there, was my initiation into sourcing suitable materials for the major projects and could be described as 'reverse gentrification' of a property. Much of that part of Glasgow consisted of tenement flats, which were more-or-less upmarket depending on the street on the hill. Those at the higher (more upmarket) end of the hill had been built with a wide range of wonderful period features, such as wood panelling, tiled cast iron fireplaces with ornate wood or marble surrounds, high-quality doors and woodwork and stained glass panels. Before being recognised for their heritage, many of these features had been ripped out in favour of 'modernisation'. These were rescued by aspiring entrepreneurs and resold cheaply, mostly in the famous Barras, a weekend market in the former Gorbals area in the east end of the city. This was a bit of a no-go area at the time and the site of the gang turf wars and the famous Ice Cream Wars in the 1980s (for the interested reader see, 'The last Godfather: the life and crimes of Arthur Thompson' by Reg McKay). The general practice surgeries were similar to a scene from 'Fort Apache the

Bronx' and Saturday nights in the Accident and Emergency Department of the Royal Infirmary was a hotbed of violent injuries. As a junior doctor in the east end of the city in the 1980s I saw a man in his mid-20s suffering, not unusually, from a knife injury to his abdomen. The chap boasted at the time that he was the last of seven siblings, all killed violently before him and that he had "died" in the same emergency department ("see me, I died here")! Reading the history books of the east end at the time this would be consistent with one of the families involved, and by dying he was referring to the cardiac arrest he had suffered a year or so earlier, from which he was resuscitated and salvaged, most unusually, by an open chest repair in the emergency department.

This Barras market sold everything from out-of-date chocolate bars to meat of dubious provenance and the rescued period features. Many of the latter were bought by young professionals who had acquired slightly less upmarket flats in the streets slightly down the hill from the high-end ones; they were essentially identical in structure but with more modest internal features. This left many of the more upmarket flats 'modernised' but already outdated with their original features reappearing in the less upmarket flats in neighbouring streets, where they fitted in perfectly well.

Travels with a staircase

When we were in Glasgow, as Baldrick would have said, I had a 'cunning plan' for the staircase in the Mill which we had just bought rather on impulse. I acquired a Victorian cast iron staircase from a remote auction in the south of Scotland and had it transported to Glasgow. It had around 15 steps, sufficient for the rise between the two levels in the Mill. Each tread was about 75cm wide and the staircase was therefore 1.5m in diameter. This staircase proved to be quite a challenge to move as each of the 15 treads weighed about 25kg. There was a space in the Mill building on one side of mill workings, approximately 2m in diameter, which I thought would be ideal for a spiral staircase, would save space and would occupy an otherwise unusable small area in the building – incorrect on all counts!

When we left Glasgow for Durham, the staircase (except for its staging which weighed much the same as a small car!) came with us in a hired van, together with various other building materials we had acquired along the way with the intention of using in the Mill. It soon became apparent that I had not done enough research and we found that the staircase would not meet the UK 2m headroom regulation at all points because of a supporting beam for the mill workings. Having been transported all the way north to The Mill it therefore returned to Durham. This must be one of the more well-travelled staircases in the country. A salutary lesson, but we were not particularly experienced at the time.

We learned a few lessons from this spiral staircase. The space required is actually greater than one might think. In addition, with an old spiral staircase separating rods may need to be fixed, possibly between the treads and definitely between the vertical handrail supports, in order to ensure that it meets the 100mm rule (that no space on a staircase should exceed 100mm). This should not apply to a modern spiral staircase, although these are extremely expensive compared to the cost of fabricating a conventional wooden stair in high-quality pine.

The space occupied by a spiral staircase is also completely unusable for anything else. In comparison, a conventional staircase also provides space beneath the rise which in the Mill was sufficient to house a dining table seating eight people with no significant headroom problems. This is shown in the photograph of that end of the lower floor after completion (see Photo 31, Page 86). My overall feeling about spiral staircases, especially old ones, is therefore that anyone should think very carefully before installing one.

The ghost in the wall

In any project several site visits will be needed before the building stage in order to assess the works required, regardless of who is to draw up the plans or dream up the internal design.

In the Mill project, the smaller lower area destined to become the kitchen consisted of a set of narrow corridors around the square base of a kiln. The window openings had been boarded up so the whole space was in complete darkness. I decided to have a look around the space to work out how best to use the area once the kiln had been cut. As I was working my way along the rather eerie, dark corridors which surrounded the kiln with my hopelessly dim torch beam, a dark-haired, dark-dressed apparition emerged from one of the window insets and then disappeared along the corridor.

I am not easily frightened but I certainly was by this! I flew out of the area, across the lower floor and up the staircase quite oblivious to the rotten treads. From my screams Gillian, who was outside the building at the time, assumed that I was no longer for this world and began in turn to scream. For those old enough to remember Donald Sutherland and the red-coated, knife-wielding dwarf in 'Don't Look Now' – the scene was not dissimilar.

A few minutes after my encounter, and I admired her honesty, the 'apparition' returned to apologise for being in the building. She turned out to be a girl who had been planning to have a tryst with her male friend who had been equally frightened; when she heard me she had hidden in the window inset. She at least had the advantage of knowing that I was there. This was possibly not the most romantic setting for a kiss and a cuddle, not least because of the floor covering consisting of a jigsaw of decaying rabbit parts, but there is no accounting for taste!

Shepherds and sheepdog trials

In the early days of the Mill project we used to stay in a bed and breakfast run by our neighbouring farmer and his wife. They invited us to a memorable day at the local sheepdog trials, which proved to be quite an interesting outing.

It began with a visit to the tea tent, which did indeed serve an amber-coloured liquid but it certainly was not tea, and indeed no tea was to be seen. Over the course of a couple of whiskies, one of the very elderly local shepherds ended up being persuaded to sit on Gillian's lap as he crooned to her. This was interrupted by a wayward sheepdog, which had herded its charge of six sheep into the tea tent and proceeded to guide them around the tent, laying waste to the tables and, to the horror of the occupants, their whiskies, before it exited (with its charges) back to the trial field.

Later on that day we went for a brief drink in the local pub, where said shepherd was sitting rather precariously on a stool at the bar until about five minutes after we arrived when he fell off. At that point we rather assumed he would be told that he had had enough to drink and politely be sent on his way.

Our assumption was a little short of the mark. He was delicately propped back up at the bar next to his umpteenth whisky, with his elbow supported in place by the back of a second stool, and so he returned to nursing his glass. We were a little worried about the poor chap as he was obviously not in a fit state to drive, and the road up the glen, however beautiful, was single track and certainly did not lend itself to inexperienced, or in this case drunk, drivers. Our farmer friend however explained to us that this was a usual occurrence, that he did not drive and that after he had drunk his fill he would walk the eight miles up the glen back to his house.

They obviously built them hardy in Sutherland in those days!

The train north to Sutherland: a bygone era in my time

It is easy sometimes to look at the past through rose-tinted spectacles. I do not think that is the case here. The Glasgow to Mallaig train route is arguably the most beautiful train journey in the United Kingdom, but close behind this must come the journey north of Durham to Inverness. Great North Eastern Railway (GNER), the franchise holder at the time, used to operate one daily train directly from Newcastle to Inverness which, if my memory serves me correctly, was the 14.47 leaving Newcastle. Onward trains from Inverness to Wick, on the north coast of the country, then stopped at remote villages along the way. It was important to catch the Inverness bound train for the comfort it offered rather than be consigned to the alternative service available. In the very early days I would travel by myself and walk to the Mill (although it was isolated, civilisation merged into total isolation over a 100m or so in the north of Scotland). Latterly we would hire a car. At the time Inverness was still rather a 'heeland toon' which seemed to have changed little since the middle of the century: frozen buttered cabbage was the order of the day and the town fell asleep after 8pm. Nowadays, Inverness is a magnificent city which has embraced the café and dining culture and has understood the desire for high-quality local produce in its eateries. It is a wonderful city where one can still sit on the banks of the River Ness sipping a glass of wine while watching an otter work its way up and down the riverbank. We still have our favourite restaurant there (Rocpool, our favourite in the UK and one I have visited some fifty times; thanks Stephen, you have welcomed Gillian, myself and our daughters warmly over twenty years, from toddlers to tequila). A Rocpool visit is now 'de rigueur' whenever we travel near Inverness, with or without any of our daughters.

Being a wily Scotsman, I would travel second class, or third had one been available. These were the times when GNER ran a dining car, the bygone era bit. At the time, first class passengers had priority access to this but once they were all seated we lesser mortals were allowed in. For around £20 in modern-day equivalent, a passenger could sit in the first class dining car and enjoy an excellent meal (remarkable given the basic kitchen facilities) as the train rattled along past the hills and lochs in the Highlands. With only one dinner service, there was no pressure to move on and I would sit there in comfort until the train arrived in Inverness.

Sadly the dining car was deemed unprofitable, a self-fulfilling prophesy once the company had banished standard class passengers from the carriage; a rather strange decision as the trains were never full so the dining car was actually a bonus for the operator. The same staff now serve the free first class offerings from the same galley kitchen so, whether travelling first or second class, there is no longer the true experience of railway dining. It was also rather sad to see the crestfallen appearance of the original dining carriage staff who had rightly taken great pride in the service they had provided and who were reduced to carrying aircraft-style food around the first class carriage.

The Boys and the Tardis

The Boys, probably the best and simplest example of finding reliable builders, carried out our largest projects – the farm complex of buildings. We had met during the early Gin Gan project, so we were fortunate in that we did not have to look too far to find them for the Farmhouse, the first of the farm projects. After becoming somewhat disenchanted with being employed at what would now be around the minimum wage, they had set out to work independently.

The Boys, Shaun and John and the tradespeople who work with them, have now become very good friends. Although on first appearance some might think they are not people one would want to meet in a graveyard on a Saturday night as they are both very strong, heavily-built gentlemen. They are in fact gentle giants, the kindest of people and immensely talented. Mark, a tradesman who works regularly with them (on roofing work in particular) is a less tall, stocky and powerfully-built Army reservist, who again is a hugely reliable, skilled worker and also a good friend. Glen, who sadly died some time ago, was at the time their primary labourer, although several other people have joined since.

Shaun is trained as a mason and John as a joiner. It was clear from our previous experiences with them from the Gin Gan and Garage projects that they could take on a wide range of tasks, avoiding the inevitable conflict in the timing of works between separate tradespeople, and that they were ideal candidates to take on the farm complex projects for us; their first major commission since they had set out independently.

I have been humbled by The Boys: their range of skills is immense and their fundamental knowledge is impressive. Academically we are probably at opposite ends of the spectrum. John told me that when he was at school anyone who worked hard

Footnotes: things that happened along the way

Photo 127 The Tardis. In this case, the reclaimed shed did not actually fit into it.

Photo 128 The Tardis ... and a few of the ultimate occupants.

was ridiculed and heaven forbid if they tried to play a musical instrument, whilst I had a privileged upbringing at public school in Scotland and then Cambridge. That meant nothing, however, when it came to building!

When The Boys ventured out to work on their own, they did so from a rather beaten-up old Ford Transit van in which they brought their various tools to site. As we were their first significant charge since they set out on their own, their range of tools was rather limited and in some cases they had to head out and hire what they needed, a process which delayed works and took someone away from the building site. As work proceeded, they seemed to invest a sizeable proportion of their income on equipment, which they brought along every day in the Ford Transit, a vehicle which became known as the Tardis.

Towards the middle of the job the Tardis seemed to carry almost every piece of small building tool known to humankind. If a small hole needed to be drilled in a two-foot thick stone wall, then a super-powerful drill and oversized diamond-tipped drill bit would magically appear from it and they would then lay pipes or cables without waiting for someone else.

As we were nearby for this project we oversaw the building works ourselves. I would regularly visit and initially caused some amusement when I turned up at their site 'bait room' in the morning dressed in my best suit ready for clinic. A 'bait room' is a colloquial north east term used to describe the room where the builders meet for lunch (bait) and breaks during the day, and which moved around our buildings as the different works were completed. When I arrived on my way to work Glen, The Boys' labourer, would have a cup of instant coffee ready to pour and a rolled Golden Virginia cigarette for me to smoke; a habit I have not been able to kick in almost 25 years and which raised more than a few eyebrows amongst my medical colleagues in the days when smoking was allowed on hospital sites. The morning planning session was amongst the most rewarding and enjoyable experiences and helped, should it ever have been needed, to build more trust in our builders.

Things are not always what they seem: a history lesson

During the Farmhouse restoration, after stripping the internal plasterwork on the gable end separating the original farmhouse part of the building from its Victorian cottage addition, we came across large stone quoins in the wall. We assumed these were simply large stones included randomly in the wall but John explained that they provided clear evidence that at some stage in the history of the building there had been an opening there as these hewn blocks of stone would never have been used in the middle of a wall. A little later, when we removed the ceilings, we discovered the remnants of an ecclesiastical quatrefoil window above the quoins. My immediate assumption was that the building must at some time have been a church. My privileged education was not much help in this situation and it was, again, John who immediately put me right without even drawing breath. The gable end of the Farmhouse building faces south whereas, as most readers will know, churches face east to west.

As described earlier, stripping off the plaster on the lower floor wall beneath the medieval window on the gable end of the Farmhouse revealed a patchwork of stone and brickwork, including evidence of the blocked up opening. This small section of wall, held together with wattle and daub, was carefully excavated to reveal the small opening. Next to this was a section of old brickwork, described as Elizabethan by our archaeologist, which had closed an opening in the middle of the wall directly below the window on the upper floor. It also contained some stonework thought to date from the Georgian era used to make good the parts around the bricked up opening. There was also some relatively modern-looking brickwork on the other end, probably dating back to the post-Victorian era, where an opening had been created between the Georgian part of the Farmhouse and the Victorian cottage.

After we had shared our discovery of the medieval window with the conservation planning officer, the grey suits (in the form of English Heritage officers) arrived, causing more than a little consternation and visions of the whole build being stopped. However, our concerns were groundless as English Heritage, or at least its regional officer, made a very positive contribution to the whole building process. The man concerned (in fact the previous Durham conservation planning officer and someone with extensive knowledge of the area) not only helped and advised us on what needed to be done but also did some research of his own. His only stipulation was that we instructed an archaeologist with expertise in old buildings to produce a report and record of the building before we covered the features up – something we had no intention of doing. The report was a superb piece of work which we cherish and was worth every penny of the £500 spent at the time. We did have to think very carefully about how best to restore these walls without damaging their heritage. We repaired the upper section of wall containing the medieval window by filling in the window space to strengthen the wall, fashioning a replacement for the missing central pillar and knitting together some stonework, again to strengthen the wall. This was finished by pointing with the same lime, cement and sand pointing mix used externally which gave an attractive final appearance. These have been cherished parts of the building in the time we have been fortunate enough to live here.

Our archaeologist dated the window to the late 13th century when this style of window, similar to those in Westminster Abbey, had been in vogue. He also found additional foundations in the services supply trench, suggesting that the original building had been significantly larger than the existing Farmhouse. Together with the reversed timbers in the upper middle room of the Georgian Farmhouse this evidence suggested that the whole building, rather than being Georgian, had been an ongoing conversion process over some 750 years. This is not perhaps found in every old building, although a little careful searching before walls are covered up may well reveal hidden historic features.

Riding shotgun on the boiler

We were fortunate that we did not have to live in a caravan during the Farmhouse project as we had funds to cover at least part of the Farmhouse works while renting nearby. Nevertheless, I did leave home and 'ride shotgun' on the Farmhouse boiler for a month or so during the later building process when the property was not fully secure but contained items, like boilers and other moveable materials that had not yet been installed, on the basis that if something was not nailed down it was liable to be stolen. The house was not habitable but I lived in a makeshift bedroom with our trusty Labrador, Hamish, as a guard dog; a triumph of hope over experience as he was yet to bark at the postie! The

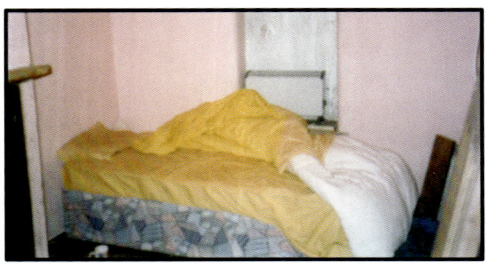

Photo 129 Farmhouse. Riding shotgun on the boiler: at least I had a heater.

Boys kindly set the room up for me, closing up the window space with insulation and with the assistance of a fan heater, a camping stove, a cheap bed and an extra warm duvet I managed to survive a number of very cold winter nights. This was an interesting experience and while rather eerie it was not actually as bad as it sounds. I did also have a chair, plate, knife and fork and a kettle, together with a plentiful supply of local carry-out meals when necessary. The rather precarious nature of the situation was best described by Mark, not someone I would take to be easily frightened as he has two tours in Afghanistan under his belt, which for politeness does not merit repetition!

Apart from the earlier case of the fireplaces, the only item stolen – by person(s) unknown and while I was camping in the unfinished house – was a cement mixer left outside by The Boys; from its tracks it was wheeled, probably with some difficulty, over the field to a waiting car. There was a little bit of poetic justice in this as the cement mixer in question was irredeemably broken and had been left, only slightly concealed, in the courtyard so that any would-be thief would take this first before attempting to break into the shed where more valuable tools were kept. Good thinking lads.

Agas, Rayburns and other stoves

Gillian has always liked Aga-type stoves, so we got one for the Farmhouse. Not a posh one to be fair, but it did look the part and provided a reasonable amount of heat. We started off with a solid-fuel Aga which was found through the local free advertising paper for a few pounds but, like one of our baths, had to be collected from a remote farm in the proverbial back end of nowhere. The Boys managed, I am not sure how, to get it into a trailer on the back of the Tardis and then installed it in the kitchen and plumbed it into a radiator and hot water cylinder. The stove provided a reasonable amount of room heat but it had to run at full burn with permanent fuel top-ups in order to produce enough heat to be able to cook properly on it and so did not avoid the need to have an electric hob and oven for real cooking. It was also a rather smoky contraption and caused no end of bother when we tried to touch up the paintwork on the kitchen walls which had acquired a thin layer of soot. We soon became tired of trips to the fuel shed in all weathers to perpetually feed it and replaced it with an old oil-burning Rayburn, which was a beast with two heads: an oil injection burner which, if it had worked, would have produced enough heat to run a central heating system; and a triple wick burner which, on the rare occasions it was not clogged up, could just about heat an oven to low roasting temperature alongside heating water and a radiator. The wicks did not in reality service both cooking and water heating and, probably unwisely and at huge expense, we persevered with it for the best part of 15 years before replacing it with a far more effective multi-fuel stove.

The modern versions of these stoves are now far more efficient than the 'antiques' we bought (admittedly for very little) and condensing boiler models are available if you have the odd £10,000 or so to purchase one. I have no experience with these. Now that we have finally replaced the Rayburn with a multi-fuel stove without a back boiler, the reduction in our heating bill has been enormous.

With the glorified eye of retrospect these were amongst our least successful purchases and by the time we had spent the money transporting, installing and commissioning them, combined with the oil they consumed, we could have paid for a few round-the-world holidays.

Oil and water baths

Apart from general rodent damage, this particular lesson follows partly from my decision, contrary to our plumber's recommendation, to use plastic rather than copper pipework in our plumbing works in the Farmhouse and partly from something we had all overlooked.

We, and unusually even The Boys, did not consider the effect of lime mortar on copper. The oil supply to our boiler was a long length of copper piping, which ran underground from the oil tank to the Farmhouse. Unthinkingly, we used unprotected narrow bore copper pipe. Combined with the effects of rodents dining out on our plastic plumbing, this oversight led to corrosion of the copper pipe by the wet lime and consequent leakage of both kerosene and water which sat underneath the floor. By the time the smell alerted us to the problem we had around 6m^2 of water and kerosene mix under the floor soaking into the fibreglass and vermiculite insulation and I spent many an unhappy hour with Henry the Hoover (see later) trying to remove it. Wet fibreglass also comes out in thousands of pieces and I spent an equally large number of hours with a straightened wire coat hanger, teasing pieces of fibreglass out rather than lifting the whole floor. I would not rate this as one of our happiest couple of weeks in the house. As described earlier, we then decided to remove the whole flooring and poured a screed over the entire Victorian extension to protect the main pipe run from hungry rodents. In retrospect, had we been aware of this potential problem we would have avoided plastic piping, ensured we used protected copper piping and perhaps used more solid insulation. Experience with this pipework, however, was relatively limited at the time.

Roadside discoveries

I have not yet gone as far as asking permission to rummage through skips at the side of the road to reclaim things headed for landfill, not out of any principle or pride but just because I did not see any with interesting contents! It really is worthwhile, however, keeping half an eye on the roadside when driving around. Apart from the wonderful oak sleepers described below, another piece of pure luck with reclaimed wood came from my travels around the county and a roadside sign advertising wood left over from dismantling a former school pavilion. This was definitely worth exploring. On immediate inspection the wood in question looked like a large pile of rubbish waiting to be burned, not helped by the fact that it was all painted a dirty shade of blue. The pile consisted of a wide variety of sizes; from 50–75mm wide long boards to posts approximately 15cm x 15cm in cross-section and around 2–3m long. Scratching a little paint off with a car key, however, exposed a dark wood which looked suspiciously like high-quality, resinous pitch pine. This was a potential gold mine, which other prospectors had missed. After a word with a local wood stripping company we bought a large amount of this, for pennies, which The Boys kindly collected in the Tardis and delivered to the company where it was put into their caustic stripping tank, for around £100, and left us with an invaluable source of boards and beams for all of the farm projects. We used a few of the 15cm cross-section posts to great effect, for both appearance and to narrow a few of the window and door openings into the barn. The narrower posts and boards were either cut on site and used to make doors, or were machined by our joinery fabricator to make excellent door jambs and architraves for the Farmhouse.

The motorised dustbin

At the time of our projects at the farm I owned Susan, a Subaru Legacy; a wonderful car not only for getting through snow in the winter but also because it had almost no electronics and therefore far fewer things that could go wrong. She carried a vast array of materials, from loose horse manure to abandoned logs from cut trees, and spent much of her time ferrying me down dirt tracks and across fields to good salmon pools on various rivers I fished, often coming home smelling rather like a fishmonger's van.

She had her moments though. One of her loads was a set of new oak railway sleepers, another roadside discovery in a small demolition yard. Reclaimed sleepers are often used to create nice regular raised beds in well-manicured gardens, but because of the weeds which grow rampant from the seeds blown from our neighbouring farmland the prospect of doing this would be rather like trying to hold back the tide. There are, however, many other uses for these marvellous lengths of wood, particularly when new and not covered with creosote or similar. The sleepers were around 2m long, 20cm wide and 15cm deep but, as delivery in this particular case was going to be extremely expensive for the relatively small amount of wood, I was faced with the task of getting them home. So, into Susan they went, sticking some 30cm out of the boot; but tied down as well as possible. I drove away cautiously, but completely oblivious to the fact that although heavy, their smooth surfaces meant that any shear force created, such as from stopping at traffic lights or in my case a roundabout, would allow the sleepers to slide over each other. The outcome was that as I stopped at said roundabout they gently slid out of the boot and onto the road! They are not the lightest of things and so after a fair degree of puffing, and a few Anglo-Saxon expletives, I finally managed to drag them back into the boot of my car and quietly prayed as I crawled the rest of the way home. There were certainly a few irate drivers who were stuck behind me at the roundabout!

Slicing these sleepers lengthways was beyond the scope of our local timber merchant, but on the manager's advice and with the kindness and generosity of a local pallet production company (Mounter and Sons, Willington, County Durham, thank you) they were sawn lengthwise into boards of different thicknesses. The Boys rounded a thick piece of wood to form a 50mm thick mantelpiece and used the remaining boards to create an assortment of finishing joinery, some shelving and a bookcase made by a local friend in return for a rather nice

Château Musar (perhaps Lebanon's best export). Compared to buying new finished oak boards of the various thicknesses required the savings were enormous, even for the relatively small amount of wood, and their final appearance was excellent. I would not have thought about doing this until I actually saw the quality and regularity of the sleepers on sale. They had also been fully seasoned and we have not had any shrinkage whatsoever over 20 years.

In terms of guttering, both the small amount of guttering still present and the hoppers and downpipes in the Mill, were all made of cast iron and were uncomplicated to replace where necessary. We sourced painted cast iron guttering parts in Glasgow and transported these in a hired van to the Mill, this achieved only a modest saving over buying new so we may have been rather overenthusiastic in terms of using reclaimed materials in this instance. We needed quite a lot of lead to replace the galvanised metal flashings and roof valleys on the Mill and were fortunate to find some code 6 (the heaviest) rolls of lead – some short but the largest almost 1m wide and 5m long – in the owners building site and left over after the works. With the help of a couple of lads we managed to get these into Susan and then with considerable effort onto a trolley in our local self-store unit. I am unsure what this did to Susan's suspension, but with the lead in the boot it gave us a good view of the horizon, but not really of the road. There is perhaps a limit to how far one can go with reclamation!

Expert opinions

We learned during our building projects to be slightly circumspect about some of the opinions given to us by experts. Apart from a couple of expensive proposals from our architect in the Mill project, which we managed to avoid with the benefit of a second opinion from a surveyor, we had a couple of other experiences which needed some thought – and in one case negotiation – before we spent our hard-earned cash.

In the first case, after we had stripped away the ceiling and exposed the medieval window in the gable end of the 'Georgian' part of the original Farmhouse, we noticed a slight lean in the stonework. The English Heritage officer also noticed this and suggested that we obtain an opinion from a structural engineer. This chap advised us that, although it did not pose any apparent immediate risk, the optimal solution would be to insert supporting stainless steel rods inside the stonework. This would probably have meant weakening the existing part of the wall initially before rebuilding it with the rods in situ in order to attempt to recreate the existing appearance. He also gave a ballpark estimate for these works as £25,000–£30,000 in today's costs. We talked this over with our architect and The Boys, who all examined the section of wall and advised that as it had stood for 750 years, the lean was likely to have been longstanding and that the proposed optimal solution was probably overkill. They offered a simpler possible solution – to leave the wall as it was, at no cost. They did, however, notice that apart from the obvious gap in the window structure, which potentially weakened the weight-bearing wall, a couple of the stone blocks had not been properly offset (i.e. they were lying directly on top of each other). Their recommendation was that they should remove these two stones and a small amount of neighbouring stonework and reinsert them so that they 'knitted' the section of wall together and would add strength without dismantling the existing stonework. They estimated this would take a couple of days' work, at a cost of a couple of hundred pounds. We took the second of these options and, while our 20-year tenure is negligible in terms of the history of the wall, there has certainly been no change over this time. This far less expensive option did little to the appearance of the original section of wall except for filling the gap in the window itself and saved an enormous amount of additional expense.

The second involved our structural work in the barn and the mezzanine area we intended to create. The beam we intended to use to support the end of the mezzanine was to be cut from one of the heavy pallet beams and to rest on padstones on either side of the building. Unfortunately this raised a bureaucratic problem. It was glaringly obvious that the beam was far stronger than modern whitewood as when two 16-stone men jumped on it when in situ there wasn't a whiff of movement and the wood itself had proved extremely difficult to cut and machine as it was so dense and hard. However, while modern whitewood has a registered strength and a beam of similar size would have been approved to support the mezzanine, the 'hardwood of unknown origin' was deemed to be unclassified and building control with advice from an unknown expert initially advised that in order to define its strength the beam would need to undergo destructive testing. This was not desperately helpful! However, with some persuasion, and support from our architect, we were allowed to use the beam on the condition that we add a supporting vertical post cut from the same wood at the foot of the staircase up to the mezzanine. This allowed us to

185

leave the beam exposed and it married fairly well with the other wood in the Barn.

We learned, therefore, that not all experts are the same and that for inexperienced home builders such as ourselves it is perfectly reasonable and potentially very valuable to garner a second opinion before following any initial advice given for a particular solution. Shy bairns get nowt!

Dead birds and bottoms

Over the years we have amassed an eclectic array of things from the sublime to the ridiculous, either in our housebuilding forays or elsewhere. These have managed to adorn most walls and flat surfaces around the Farmhouse and the Barn; the later Barn Conversion meant that we had to rehome some of these. Two items in particular have continued to bring amusement years after the event.

On a visit to the lovely village of Helmsdale, famed and still celebrated for the late Dame Barbara Cartland, we stopped at one of the two fish and chip shops located on either side of the main road (how a village of very few people and only a limited number of visitors supports two fish and chip shops I really do not know). Anyway, we ventured into one, which it is perhaps a little unfair to describe as a fish and chip shop as, alongside fish and chips, its menu included a range of local seafood fresh off the few boats which still operated out of the harbour. On our first visit we sat down to the same dish each, toddlers in tow. This was a pint of local prawns followed by a fish medley, a marvellous dish of haddock and a range of shellfish. Our hopes were slightly dashed when the waitress came back after taking our order to say that the crayfish were not at their best, but were lifted again when she said she would include half a lobster each instead. The wine was of one common brand, red or white, and everything was extremely simple but good. The restaurant also had the most bizarre mix of assorted memorabilia for sale, including a taxidermy buzzard, one of my favourite birds, so home it came with us to sit in the barn. Shortly afterwards I acquired, through an acquaintance, a rather tasteful coilwork pottery outline of a female figure, shoulders and behind, held together by two leather straps; which also found its home in the barn.

When we converted the Barn we kept our more treasured possessions, including these two items. The buzzard was housed on an upstairs hall chest and the bust on the wall in our central sitting room. All was well until one of our daughters returned from a holiday and came down one morning in extremely high dudgeon to say "There's

Footnotes: things that happened along the way

Photo 130 Our offencing buzzard.

Photo 131 Our even more offending bottom.

a bl**** dead bird outside my bedroom and a f****** arse on the wall". Beatrice was, and still is, the most softly spoken of girls and her outburst simply left us in tears of laughter, which only made her even more angry. We still laugh about it today, much to her chagrin. This story has been recounted numerous times mostly to the enjoyment of those present. Not quite so however on one occasion when Gillian was in Glasgow visiting our eldest daughter. They planned to go out for a meal in their favourite (unlicensed) restaurant but unfortunately it was full. Bottles clinking in bags they trudged down the road to their next favourite, a rather posh (but licensed) small Indian eatery, already feeling slightly embarrassed by their plastic bag of rattling bottles. The restaurant was quite quiet with only two of the eight or so tables occupied; one by a group of rather erudite-looking, possibly academic, Indian gentlemen and another by a rather genteel-looking family. Gillian began to recount the story and got to the final sentence when for some reason all went quiet in the restaurant and all that could be heard was the punchline "f****** bottom on the wall". Between the bottles and the bottom they felt they had brought enough shame upon themselves, hastily finished their (rather good) dinner and exited not to return for long enough until hopefully the event had been forgotten by the taken aback waiter and the other diners.

Auctions, flying Chesterfields and other tales

One option which many people overlook outside of internet auction websites such as eBay is local auctions. I am always surprised by the number of people who have never attended one. I think this is possibly because in many cases it has not crossed their minds, although the clutch of television antiques and auctions programmes may have brought these more into focus. Auctions can be a good source of materials and I also find them enormous fun. Local dedicated auctions for building materials come up from time to time but general auctions can still be valuable. The whole process has also been made easier by online catalogues and bidding but, like betting on the horses, there is no substitute for being at the event in person. Old benches and cast iron radiators are not infrequent finds as occasionally are materials left over from other builds.

One caveat however is the temptation to go 'off-piste'. As we were going to need a considerable amount of additional furniture because of the sheer size of our buildings, it was tempting to look at furniture. Although perhaps a little eccentric in choice, we did supplement our home with lots bought at these auctions. The cost of old 'brown' furniture had fallen dramatically at the time, partly because of a shift in preference in the UK towards a more modernist approach to furnishing and partly because of the fall in the value of the dollar against the pound at the time, hitting the American market where much of the older traditional pieces of British furniture often went. Most soft furnishings have virtually no auction value so can be picked up for pennies.

Our relatively local, and occasionally peripatetic, auctioneer is a chap called Jim Railton, a great fun and very effective operator, used to hold biannual two-day events in our local castle, a mile away, so delivery costs were negligible. These auctions comprised some 2,000 lots ranging from broken chairs going for 50p to Lowry drawings selling for many thousands of pounds. They were an institution at the time and we spent many happy days over the years in a rather cold, large hall used as a temporary auction room nursing cups of coffee and watching events unfold.

I have a slight penchant for leather (of the furniture variety that is) and we also acquired, again for pennies, a three seater oxblood-coloured Chesterfield sofa for the 'snug' end of the barn. It looked rather lonely on its own so in order to find it some friends I found a similar style fireside armchair some 20 miles away in Newcastle, which I carried back on the roof of my car. Driving fairly cautiously on the A1 the chair nevertheless broke away from its moorings and flew off, clipping the car in front of me, fortunately causing no damage, and then rolled down a 30-foot incline where it came to rest. The small car in front was actually brand new and had just been collected by an elderly couple who emerged, ashen-faced, from their vehicle after we stopped. I drove to the next roundabout, turned around and returned to recover said chair. At that point a police car arrived, blues and twos flashing, and, although I was fully expecting to be cautioned for reckless driving, the policeman very kindly helped me put the chair back onto the roof and tie it down (this included giving me a knife to cut the rope, which must have broken every police regulation). He then, very courteously, suggested I stop at the next garage to buy some more rope. I did so and managed to limp back home without further incident. Remarkably, the chair suffered no damage other than a scuff on one wing, which it has happily sported as a war wound. What a nice experience with North East Police.

Shortly after buying these two Chesterfields I stumbled over a rather nice club seat of identical style, again in the same matching oxblood colour and in excellent condition on our back lane, a nirvana for fly-tippers. No shame, no gain!

Henry the Hoover

At this point I need to acknowledge 'Henry the Hoover'. In fact he (or she) is a Reddy Vac, but has always been known as 'Henry' and was acquired at a local car boot sale for a couple of pounds when we were carrying out the first of the farm projects. He is now 21 years old, plus whatever age he was at the time we bought him (and I don't think he was a spring chicken at the time). He has been used exclusively for 'dirty' cleaning throughout the farm projects and over the last 20 years or so and has hoovered up everything: rat droppings, sawdust, sand from internal sandblasting, lightly vacuuming exposed stone walls to remove sand and salt as the walls dried out, removing kerosene and water-soaked vermiculite insulation from underneath flooring, ash overflow blocking the back of woodburning stoves and all of the jobs that one wouldn't want to consign to a

Photo 132 Henry the Hoover.

'clean' vacuum. He has also, and I am sure extremely dangerously, removed the final few gallons of water from a plunge pool (not to be recommended at home). However, probably his worst case of indigestion came when I inadvertently used him to suck up the remnants of a mouse deposited in an outhouse by one of our ferocious cats. Together with the neighbouring pieces of straw, this created a rather strong organic glue, which tested even my most robust rodding system. 'Henry' has been another of the best investments we made back in the day and still works as well as he did when we first bought him, even if he has lost a castor and limps along a bit, looks a little bedraggled and has needed the occasional rodding to remove assorted things that have managed to clog up his suction pipe. A very wise investment and one which I would thoroughly recommend to any aspiring builder as part of the equipment armamentarium. Rather like several of the building solutions themselves, simple and inexpensive can be best – Henry fitted the bill on both counts.

Country living

One thing which applies to a fair number of dilapidated buildings, particularly ones in the countryside, is that they are often located on back roads (rather than thoroughfares)

Photo 133 Country living. When the road floods, it floods.

Photo 134 Country living. Bless the police for trying to divert cars, but this queue goes back 200m despite 20 cars in our driveway.

Photo 135 Country living. Definitely not a lane for a rear-wheel drive.

Footnotes: things that happened along the way

Photo 136 Country living. At the end of the day; many sunsets like this with no light pollution.

Photo 137 Country living. Kune Kune pigs; nightmares on legs.

Photo 138 Country living. Free-wandering chickens: fun, productive and friendly. Much safer than geese!

which are not well suited for extreme weather conditions and may not be cleared or gritted by local councils. Our farm complex fits this description perfectly, it also lies between two hills; one gentle but long with a couple of steep dips along a flat section of road, and the other both steep and long. Both are prone to icing, making it difficult to get up either slope, and the dips in the flat section are also prone to deep impassable puddles in wet weather or from snow melt. At its worst, when the main road flooded the local police, I am sure with the best of intentions, directed all traffic down our lane to bypass the flooded main road, only for the drivers to find that the road hollow had grown from a modest pool into a pond two feet deep. As a result no traffic could pass, and we ended up with 20 or so cars parked in various places around the farm and a further 50 or so stuck on the impassable lane.

When we first moved into the Farmhouse we were unfortunate to have a large rear-wheel drive estate car (useful at the time for our three young daughters and an oversized Labrador the size of a Shetland pony). However, we had major problems with this as it had no hold on icy roads and getting up the steep hill on our lane in such conditions was out of the question. On a few occasions, in particularly freezing snowy weather, the only way of reaching our local village and the main road was to put the car into reverse (i.e. effectively converting it into a front-wheel drive) to travel the kilometre or so to the bottom of the lane). We left it there for several days and ferried ourselves to and from it with our second, smaller, four-wheel drive car which just about managed to get through the snow and ice. We have spent many hours proffering tea to unfortunate (and often unwise) people who have tried to drive up the particularly steep hill immediately past the building in icy weather, often in rear-wheel drive cars, or those who have ended up in deep roadside ditches because either they or an oncoming driver did not know how to drive on a single-track road.

This phenomenon is not restricted to mid-winter. On one, now locally famed, occasion we hosted a party for our eldest daughter in late autumn immediately after a rather wet spell. Because of the number of, in some cases exceedingly expensive, cars these were all directed into a paddock to park. Unfortunately, at the end of the party almost none of the rear-wheel drive cars managed to get across the sodden field but ground themselves into the mud in their attempts to escape. Perhaps the highlight of the party, at least for the children, was watching the spectacle after the smaller front-wheel drive cars had left, as Mummy or Daddy's rear-wheel pride and joy was unceremoniously towed out of the mud to safety by David, our neighbouring farmer, and his JCB. The children could not have been more excited, which is more than can be said for their parents. I suggested selling the rear-wheel drive car, however loved it was, and buying a sensible four-wheel drive.

I'm just popping out for a chicken ... and a few other things

One of the benefits of country living is the opportunity to experience a little of 'The Good Life' (not quite the 1975 version in Surbiton for those who

Photo 139 Country living. A large pond; a good and inexpensive groundworks investment. We did not add fish so have attracted many birds.

Photo 140 Country living. A year 20 overhaul. It took about six months for the wild grasses to re-establish.

can remember the TV series). Over time, we collected various forms of poultry in cat baskets or straw-lined cardboard boxes. The occasional flighty duck did, however, manage to escape from its box and flap around the car to the shock of neighbouring drivers waiting at traffic lights; a real double take if I ever saw one. The range of poultry and, the rats and foxes that unfortunately come with them, did not prevent us extending our collection in addition to the usual array of dogs and cats, goldfish, budgerigars, hamsters and a psychotic parrot which came with children. These additions ranged from the common to the more obscure, including rare breed ducks and geese, rheas, and 'miniature' pigs.

Chickens are fun and productive, provided that one is happy to sacrifice any form of flowerbed or vegetable patch which can be used for a dust bath. Our foray began with a pair of Saxony ducks, bought in mid-winter. Sadly, one wandered out into the snow and, to quote the famous words from the tragic Antarctic expedition "I am just going outside and may be some time". Although we managed to retrieve it just about alive, sadly it succumbed after a day sitting in the Christmas turkey box by the Aga. Perhaps the box may have given it a premonition about its future. These were followed by a range of other ducks, from Indian Runners, through Aylesbury to Muscovy ducks (actually a form of goose), which although not the most attractive of birds are wonderful characters and were great fun to watch taking off and landing like B52 bombers for their morning constitution around the farm. At our peak we had a mix of fifty of so assorted chickens and ducks, supplemented by two geese which are certainly good guard animals; so good in fact that they successfully intimidate their owners. Possibly as a twist of fate these were rehomed by a couple of lads who came to pick up an old dingy I no longer used (except perhaps in reserve to navigate our lake on the lane in torrential rain!).

Photo 141 Country living. We are delighted when the first pair of greylags appeared. For us the first day of spring.

Photo 142 Country living. We now have eight returning pairs (some shown). A bit noisy but they fall asleep before we do.

Astonishingly, they took a liking to the geese and asked me if they were for sale. Not going to miss an opportunity I suggested that if they could catch them they were welcome to take them home. Armed with their torches they went on a hunt, and wherever they ended, on the table or as pets, they no longer terrorise us or our daughters when we arrive home.

Then came a trio of rheas, small ostrich-like birds and possibly the only animals we have encountered which have less sense than a chicken. After this came possibly our biggest mistake, miniature pigs in the form of a pair of Kune Kune, which are not exactly that small. As anyone who has tried to pull an unwilling small pig escaping down the road on a dog lead will attest, these animals are wilful, extremely strong and really quite intelligent. They learned to escape through their electric fence, which seemed to give them only a frisson of excitement, lift a gate off its pin hinges and break into the feed house to gorge themselves on a bin of pig food. They would then lie, quasi-motionless, emitting enough methane to constitute an explosion hazard until they were in a fit state to move. Rehomed to a kindly but rather eccentric local barrister with an interest in breeding these animals, they were possibly our greatest source of frustration. We now stick to chickens.

The above however is not the end of the poultry story. On one occasion, one of our chickens had been attacked by a local dog and had been torn open from body to wing. We had wrongly assumed this would herald the demise of the bird, but in an attempt to save it I went to my local hospital and asked the sister in charge of the Accident and Emergency Department if I could possibly have some heavy-duty catgut sutures to sew up a chicken. Assuming this was for eating and to tie the back end to keep chestnut stuffing in, she asked me what grade we wanted and was, I think, a little taken aback when I explained that the chicken in question was a live one. Despite my limited experience as a surgeon (none if the truth be told) the said chicken lasted for a further two years, so perhaps my surgical skills are not as bad as all that!

Photo 143 Country living. The roe deer in the woods are rather elusive: our eco-friendly reclaimed iron deer moved down from the north.

Memorable holidays

I have described many successes and a few failures of our own making in this book. Fate and Acts of God, however, will play a part or go alongside any self-build project. This story came alongside. During and for a long time after the Farmhouse build, we went up to the Mill at various times of the year, but most often around Christmas time. On one of our later visits before the Farmhouse was complete we went on a slightly ill-fated one.

We had decided to spend a week in the Mill to see in the New Year. The trip north started inauspiciously when our large Labrador, who seemed to live in a perpetual state of worry at being left behind, jumped into my Subaru Legacy when we had left the back door open and decided to eat a good fraction of the back seat. Well, it was an old car so we did not mind too much. The northbound journey passed with few events apart from one of our young children, as they were then, being sick at our breakfast stop which affectionately became known as the chuck-up inn (nothing to do with their food) as it seemed to become a regular point for travel sickness. Arriving otherwise intact in Sutherland, apart from the inevitable smells which come with a large dog (Mummy, Hamish is (dis)gusting), we passed an excellent New Year capped by a ceilidh in the local community hall. The return journey was less successful. It began with snow warnings and with a little (fortunate) insight we chose to avoid the A9 and take a back route through Aberdeenshire and down the east coast. Albeit a bit late we managed to get through this road, probably one of the last cars to do so and passed, on a particularly steep hill, a poor Mini which had skidded to a halt midway up. By that time, the A9 had been closed to traffic.

Reaching home in the dark a couple of hours later than planned, we arrived in a biblical downpour to find that the lock to the one door we had left unbolted had jammed. No way in! With a little help from our neighbouring farmer and a crowbar, we managed to get in and open a door from inside. So far so good. We then tucked our children into bed and headed the same way ourselves only, to find that a cat had relieved itself all over our bed, presumably in revenge for the lack of attention during our absence, at which stage we gave up, managed to find another duvet and moved to another room. I was woken in the night by the sound of heavy rain but went back to sleep. It was only in the morning that we discovered that the heavy rain had actually been water pouring down a wall inside the building because of a ballcock which had sheared off during the night (contrary to the impression I may have given in this book, this was actually new and not reclaimed). To add insult to injury it had run directly through a cupboard, soaking and staining all of the linen, which my wife had just washed. This was not one of our best 24 hours.

Amongst the other problems, there are few means to prevent a revengeful cat from urinating on your bed, other than obsessively ensuring that all doors are shut; but anyone who has owned a cat will know that they are the wiliest animals and can slink through the narrowest of spaces unseen.

To conclude:

Cats no less liquid than their shadows
Offer no angles to the wind.
They slip, diminished, neat through loopholes
Less than themselves; will not be pinned

from Arthur Tessimond, 'Cats'

A few finishing words

I would not dare to claim exclusive rights to imagination and some of the ideas contained in this book might seem simple common sense. The examples and solutions described are exactly that, examples, but I think they show that with a combination of imagination, some risk taking, lateral thinking, tenacious perseverance and some opportunism, it is possible to defy some of the received wisdom and greatly reduce the common cost estimates of restoring an old building or wreck. This can also achieve as good or, in our experience, a better result than following conventional approaches. Depending on individual attitude to risk, this also adds to the fun and richness of the experience, which most would-be restorers will carry with them for the rest of their lives. However, despite all the joy and laughter, expect a few tears along the way.

Photo 144 Postscript 1994. Memories of Sutherland where it all began in earnest. The first of the major projects began because of our love of salmon fishing and a vision to build a home close to the river.

Happy building and may the force be with you!

Glossary

The terms described briefly in the glossary below are defined only in terms of how they are used in this book and how they relate to our buildings or building methods.

A-frames	An A-frame is a basic structure designed to bear a load, in this book, a roof attached at the top and with a lower crossbeam forming an uppercase letter 'A' shape.
Bulkhead	As described in this book, a bulkhead is an upright partition that separates compartments of the building extending from the floor to the apex of the roof.
Casein-based paints	Casein is an ancient type of milk-based, water-soluble, opaque paint which is highly breathable and, in the case of an interior side of a solid stone wall built using lime, helps regulate humidity by allowing water vapour to pass through it.
Coefficient of performance (COP)	The ratio of energy delivered to energy used by, in our cases, a heat pump. The coefficient of performance or COP of a heat pump is the ratio of useful heating or cooling provided to the work (energy) required to run the system.
Consumer unit	A component in the electrical supply at the point it enters a domestic property, which contains devices such as a switch and circuit-breakers.
Door jamb	One of the two vertical posts on either side of an opening into which a door fits.
Double-bunded	A bunded tank is essentially a tank within a tank. The term is used in this book to describe tanks holding oil for domestic use. The oil is held in the inner tank or 'skin' and the outer skin is a secondary tank that offers protection.
Dry-lining	A form of cladding, in this book insulated and used to line the internal walls; plasterboard is attached to batons fixed to the wall over a permeable or impermeable membrane to create a smooth surface ready to cover with a layer of plaster. In some cases the plasterboard is painted directly after taping the joints.

Elevations	In architecture: an elevation is the front, back or side of a building; or a drawing of one of these.
Feed-in tariff	A payment made for the generation of electricity, by unit of energy produced (kW), through the use of 'renewable energy' methods, in this book solar or wind energy.
Fused spur	A switch that is used to isolate electrical equipment from the mains supply.
Gin gan	A gin gan is a horse-drawn mill building with stone pillars separating open spaces, but roofed and containing millstones and supporting structures. In our cases these were hexagonal in shape and connected to barns which would have contained additional milling equipment. Most were built in England in the late 18th and early 19th centuries.
Gravity-fed system	Gravity feed is the use of earth's gravity to move something (usually a liquid) from one place to another. In this book it refers to central heating systems and incorporates a tank located above the boiler and hot water tank, fed by the mains water supply, to allow water to pass into the heating system.
In-principle approval	As used in this book, the term refers to the approval given from an initial approach to the planning authorities that the proposed building works do not raise any major objections, prior to submitting a detailed application for planning (and Listed building where this applies) permissions for the build.
Inverter	An electronic device or part of a circuit that changes direct current to alternating current.
Lath and plaster	An interior wall construction technique that typically predates the 1940s. Strips of wood (laths), typically 4" long and 1" wide, nailed directly to the open walls, used as the base to apply layers of wet plaster.
Ledge and brace door	A traditional, sturdy door constructed with the number of vertical wood boards needed to form the width of the door, held together by horizontal wooden boards (ledges) and diagonal ones (braces).

Glossary

Padstone	A block made from stone or concrete that is used to distribute a point load evenly into a structure, in this book the ends of beams and lintels built into a wall.
Permitted development rights	A national grant of planning permission which allows certain building works and changes of use to be carried out without the need for a planning application. Permitted development rights would not normally apply to Listed buildings.
Purlin	A horizontal beam along the length of a roof, resting on principal rafters and supporting the common rafters or boards.
Road planings	Road planings are produced when the surface layer of a tarmac road or footpath is removed. These can be used as a base or surface for a driveway or other area of the building grounds.
Scrolled	In this book, the process of scrolling describes shaping the length of floor board which lies against an irregular wall, such as one constructed of stone.
Second-fix joinery	All of the finishing joinery work in building. Second-fix joinery tasks include internal doors, architraves to door linings, skirting boards, staircase components and low-level boxings to run plumbing pipework and/or electrical cabling.
Single/Multi-head split system	In this book a single-split system for an air-source heat pump consists of an indoor unit and an outdoor unit connected by pipe, a one-room solution which is optimal for targeted heating or air conditioning. A multi-split system may have several (normally up to five) indoor units connected to one outdoor unit. Multi-split systems are suitable for multiple rooms. A major difference is that the single-split system is not disruptive to install and does not, therefore, need to be installed during initial building works, whereas the multi-split system requires more complex installation.

Solum/Subsolum	The solum and subsolum in soil science consist of the surface earth and material beneath the surface (e.g. earth or clay), respectively.
Tanking	Also referred to as below-ground waterproofing – involves the application of a waterproof barrier to the walls, previously bitumen and now a waterproof cement-like render to a below-ground wall structure.
Tee hinge/T hinge	A traditional general-purpose hinge, normally made from steel and coated with a protective finish, with a frame fixing flap, a straight hinge and a tapering horizontal strap with screw fixings along its length, typically used on doors such as those made of a ledge and brace construction which cannot be furnished with the modern hinges fixed to the side of the door.
Turnbuckle	A stretching screw or bottlescrew is a device for adjusting the tension or length of ropes, cables, tie rods, and other tensioning systems, in this book for wires forming the sides of staircases and balconies.
U-value	Thermal transmittance is the rate of transfer of heat through matter such as a roof or wall. The thermal transmittance of a material or an assembly is expressed as a U-value.
Wattle and daub	A material formerly or traditionally used in building walls, consisting of a network of interwoven sticks and twigs covered with mud or clay.
Whitewood	A term loosely used to describe inexpensive basic grade wood (typically spruce) used as described in this book for flooring.
Zone valve	A type of valve that is used to regulate how much water flows through a closed heating system which, in the context of this book, describes a valve within a central heating system used to separate areas of a building and programmed to open separately.

List of photographs

Photo 1 On the way to the Mill. Gateway to the real north: the Dornoch Firth and Kyle of Sutherland where sheep take over from humans along the A836, a circuitous route back to the northbound A9 over the Struie Hill. On my last trip to and from Inverness to the Struie turning, around 12 miles away, I counted eight red kites and countless buzzards.

Photo 2 Mill. Front of the original building.

Photo 3 Mill. After removal of earth covering the windows (and the tree).

Photo 4 Mill. The Mill wheel required only a tidy and a spray with a proprietary preservative.

Photo 5 Mill. Appearance of internal wall with roofing timbers.

Photo 6 Mill. Detail of the rear of the Mill building.

Photo 7 Mill. Upper floor, doors to smaller grain drying area: iron rails support the perforated cast iron tiles (see Photos 26–29).

Photo 8 Mill. Upper floor. Opposite direction. Rather daunting.

Photo 9 Mill. Larger part of the upper floor facing away from the smaller area, looking towards an east window.

Photo 10 Mill. Minor repair and cleaning works restored the mill workings.

Photo 11 Mill. Door to the lower smaller area to kiln with a surrounding 'corridor'.

Photo 12 Mill. Kiln cut and part-finished second 'corridor' area used to house the boiler.

Photo 13 Mill. Original mill workings obscured by a damaged façade.

Photo 14 Mill. Mill workings after removing wooden façade and cleaning.

Photo 15 Mill. A tranquil spot to enjoy a glass of Chablis and the sounds of running water and the occasional screech of a buzzard overhead.

Photo 16 Mill. A picture-postcard view at the end of the garden. Location, location etc.

Photo 17 The North East has a host of 'don't miss' venues from sea to dales. Tynemouth market is great for browsing and street food.

Photo 18 Farm. The farm buildings had to be stripped out internally, but the building fabric was sound apart from the chimneys (here).

Photo 19 Farm. Aerial photograph of the farm (1964) showing, amongst others, the additional structure in the courtyard against the barn.

Photo 20 Farm. Recent aerial photograph of the farm from a slightly different angle.

Photo 21 Farm. Early view of the 'Victorian' end of Farmhouse.

Photo 22 Farm. Early view of the back of the barn showing the gin gan.

Photo 23 Farm. Early view of Victorian extension showing three access doors.

Photo 24 Farm. The Marley tiles are not out of place against the Barn pan tiles.

Photo 25 Farm. View to the Byre. Building sites make great playgrounds.

Photo 26 Mill. Upper floor to the smaller area with the heavy iron doors removed showing the entrance to the drying room.

Photo 27 Mill. Completed upper corridor with beech flooring into the new bedroom created in the upper kiln drying/malting room.

Photo 28 Mill. Finished bedroom in upper part of kiln. New windows, rooflights, cupola and light Junckers® flooring all add to the lightness.

Photo 29 Mill. The star-shaped beam structure supporting the kiln ceiling, stained by years of kiln smoke.

Photo 30 Mill. Finished open-plan area looking towards the fuel hopper.

Photo 31 Mill. Completed open-plan area looking towards the mill workings.

Photo 32 Mill. View to mill workings with open plan area and staircase detail.

Photo 33 Mill. Kitchen with cut kiln and supporting pillars for upper floor. The central island is

Old Buildings: Conversion and Restoration

inside the former kiln.

Photo 34 Mill. The ebony-stained timbers contrast the Caithness flagstone flooring and white interior.

Photo 35 Mill. As a break from emulsion, a few simple DIY stencils worked well.

Photo 36 Mill. Front view. The copper cupola and has aged well over the years.

Photo 37 Mill. Early appearance. Back view from the back.

Photo 38 Mill. Final appearance. Back view

Photo 39 Mill. A find on our quest for materials, by Lairg, Sutherland. That far north, 'by Lairg' covers a 30-mile radius!

Photo 40 Farm. The original Farmhouse walls needed extensive repointing: the existing build structure dictates how this can be done.

Photo 41 Farm. The same wall 20 years later: no internal stonework deterioration or dampness.

Photo 42 Farm. The whiteness of the aged lime mortar (the small section below right) stands out more against the stonework than the wet mix containing just a small amount of cement

Photo 43 Farmhouse. More of architectural interest: the stone quoins uncovered in the gable end of the former Farmhouse.

Photo 44 Farmhouse. Initial appearance of the medieval window found above the existing ceiling.

Photo 45 Farmhouse. The medieval window after restoration and masonry work to strengthen the gap in the window.

Photo 46 Farmhouse. Some of the other back of the window remained (plastered over): careful sandblasting avoided damaging the stonework.

Photo 47 Farmhouse. Part of the lower gable end wall showing evidence of a closed up section.

Photo 48 Farmhouse. Careful removal of the wall covering revealed the opening. Easily missed.

Photo 49 Farmhouse. Repair to the stonework (same 1:2:6 cement, lime and sharp sand mix) and exposure of the woodwork in the opening.

Photo 50 Farmhouse. The whole length of the central lower gable end of the Georgian section.

Photo 51 Farmhouse. 750 years of history in a wall. The lower gable end shows the progression from medieval times to the post-Victorian era.

Photo 52 Farmhouse. The uncovered walls revealed another inset which had been closed with stonework and plaster.

Photo 53 Farmhouse. One of many small features found by careful inspection of exposed walls.

Photo 54 Farmhouse. Middle upper room giving a better view of the original older roofing timbers.

Photo 55 Farmhouse. Middle upper room wall with covering removed shows the lintel from the original opening and stonework.

Photo 56 Farmhouse. Middle upper room. Finished roof timbers showing pegged joints and slightly convex pegged arms of the A-frame.

Photo 57 Farmhouse. Middle upper room. Finished gable end and exposed timbers with the vaulted ceiling.

Photo 58 Farmhouse. Gable end upper room with fireplace stolen. Initial appearance.

Photo 59 Fireplace. Gable end upper room after stripping out before dry-lining, revealing the original chimney breast.

Photo 60 Farmhouse. Same room with finished chimney breast.

Photo 61 Farmhouse. Middle upper room: the more recent stonework was unremarkable and rendered, leaving the chimney breast exposed.

Photo 62 Farmhouse. The fireplace area in a lower, bedroom in the Georgian end. Initial appearance.

Photo 63 Farmhouse. The same fireplace in a lower room sandblasted and repointed.

Photo 64 Farmhouse. Above the brickwork, very old notched timbers suggest rebuilding dating

Lists of plans and photographs

back centuries. Left exposed.

Photo 65 Farmhouse. Central room with exposed upper floorboards: the offending modern fireplace before removal.

Photo 66 Farmhouse. Excavated fireplace in central room. 'New' stove with back boiler and blackened stonework from a few centuries ago.

Photo 67 Farmhouse. Winter warmth. The stove and its two radiators are sufficient to heat the whole core of the house without burning oil.

Photo 68 Farmhouse. A vintage Rayburn drank oil but at least kept bottoms warm.

Photo 69 Farmhouse. Replacing a thirsty wick burner oil Rayburn with a woodburner provided more heat at a fraction of the cost.

Photo 70 Farmhouse. The most complicated replacement Georgian sashes produced by our fabricator: indistinguishable from the originals.

Photo 71 Farmhouse. The Victorian cottage end. Retaining two different window styles was essential to show the history of the building.

Photo 72 Farmhouse. Two joists added after removing a staircase: the reclaimed additions are indistinguishable from the originals.

Photo 73 Farmhouse. Underside of the middle Farmhouse upper room left as the exposed ceiling to the lower room

Photo 74 Farmhouse. Damaged sections of the underfloor: old repair on right (leather strap) and same style repair (stained batten) on left.

Photo 75 Farmhouse. The cleaned upper surfaces: not perfect but the damage marks do not detract from their overall the appearance.

Photo 76 Farmhouse. The original Victorian bedroom fireplace and surround.

Photo 77 Farmhouse. Sandblasting combined with a great deal of hard work stripped this fireplace back to the original.

Photo 78 Farmhouse The initial appearance of the Georgian staircase, at first sight beyond redemption but well worth looking at more closely.

Photo 77 Farmhouse. Sandblasting and scraping returned this fireplace to the original.

Photo 78 Farmhouse. The initial appearance of the Georgian staircase, at first sight beyond redemption, worth looking at more closely.

Photo 79 Farmhouse. Same staircase after sandblasting. The wood indents made any alternative cleaning method pointless.

Photo 80 Farmhouse. The staircase finished and sealed; replacement was unnecessary and the repair provided a good final result.

Photo 81 This young lady came from a small island off Guadeloupe via foot, ferry, car, taxi, plane, second taxi and plane, then car to Durham; all with a toddler with incipient chickenpox and Gillian pregnant with twins!

Photo 82 Farmhouse. Lower Victorian room. Original plasterboard ceiling removed: joists and underside of upper floorboards exposed.

Photo 83 Farmhouse. The final appearance of the lower Victorian room, with exposed joists and timbers.

Photo 84 Farmhouse. The same range, cleaned and finished with Zeebrite®. A wonderful relic from Victorian days.

Photo 85 Farmhouse. An original door. Slightly moth-eaten, but fine in an old Farmhouse.

Photo 86 Farmhouse. A new door with reclaimed sandblasted hinges and latch, very close to the original.

Photo 87 Farmhouse. An external door: there is absolutely no need to replace many original doors.

Photo 88 Farmhouse. Eye off the ball. The reclaimed wood above clashes dreadfully.

Old Buildings: Conversion and Restoration

Photo 89 Farmhouse. Reclaimed pitch pine panels used for the doors on reclaimed carcasses.

Photo 90 Farmhouse. The final appearance of the kitchen; rather busy but it works.

Photo 91 Farmhouse. Reusing the existing bathroom fittings posed no problem and looks fine.

Photo 92 Farmhouse. After sanding our behinds, this bath found its place back outside: a case of over-zealous reclamation!

Photo 93 Farmhouse. Finished appearance of the front of the whole Farmhouse.

Photo 94 Farmhouse. Home and Happy, 21 April 2000.

Photo 95 Barn. Attractive roofing A-frames and purlins. Thin, highly-insulating materials used to leave the purlins exposed.

Photo 96 Barn. Under construction: the reclaimed modern hardwood structural beams. Not old but fitted in well and cost pennies.

Photo 97 Barn. The reclaimed council radiator without its new cover and partly laid Douglas Fir flooring.

Photo 98 Barn. Finished Douglas Fir flooring (reclaimed old floor joists) imperfections fitted with overall look; sealed with Bourne Seal®.

Photo 99 Barn. Finished Barn looking away from the Farmhouse. Open-plan 'snug' with the dining and seating areas.

Photo 100 Barn. Finished Barn looking towards Farmhouse with reclaimed Douglas Fir flooring and exposed roof timbers.

Photo 101 Barn. The finished Barn: view from the new mezzanine.

Photo 102 Barn. The arrangement of the gin gan timbers is similar to the kiln end of the Mill: similar pyramidal roof shapes.

Photo 103 Barn. Finished view of the gin gan. Left much alone other than adding a floor, windows and ceiling boards.

Photo 104 Barn. Outside north side: minimal change from the 1964 aerial photograph (Photo 18) or early appearance (Photo 22).

Photo 105 Byre The outside of the original Byre facing into the courtyard.

Photo 106 Byre. The same wall after cleaning and repointing, with replica windows matching the damaged originals.

Photo 107 Byre. A 'bait break'. The frame for the bathroom has been erected and the animal stalls are still in place, some dividers to be kept.

Photo 108 Byre. Finished. Shows roof timbers, living and kitchen area (demarcated by animal stall divider), bathroom and door to bedroom.

Photo 109 Farmhouse Extension (right). Closed-up redundant entrance, outbuilding doors and windows painted heritage blue. In keeping with the original, the extension windows and entrance doors are finished in antique pine (Sadolin®).

Photo 110 Farmhouse Extension. Front view of the Extension.

Photo 111 Farmhouse Extension. The corrugated roofing blends in colour with the Barn roof as in the early aerial photograph but, as it was originally, the roofing type is quite different to the main building. Not done to save money, but this construction was also considerably less expensive.

Photo 112 Farmhouse Extension. Internal view of the Farmhouse Extension. As it is south-facing, the room attracts heat and is warm even on cold sunny days. It has become one of the most used rooms in the Farmhouse.

Photo 113 Barn Conversion. Old closure in a wall used to support the planning application.

Photo 114 Barn Conversion. New mezzanine, view towards Farmhouse. Glass panels open up the building avoiding wooden spindles.

Photo 115 Barn Conversion. Panoramic view from the mezzanine over the open-plan area.

Photo 116 Barn Conversion. Same area, view away from Farmhouse with new entrance.

Lists of plans and photographs

Photo 117 Barn Conversion. Towards the Farmhouse: open view through the building.
Photo 118 Barn Conversion. Little changed in the gin gan other than to relocate a stove.
Photo 119 Barn Conversion. A slightly risky choice of door to add natural light in an old building but it seemed to work.
Photo 120 Barn Conversion. A lower bedroom, same wall opening and floor and the closed up entrance through to Farmhouse.
Photo 121 Barn Conversion. The difference in quality and appearance did not seem to justify the high cost of specialist suppliers.
Photo 122 Barn and Barn Conversion. Old fan-assisted radiators. Covers made on site: rapid heat and saved wall space.
Photo 123 Barn Conversion. Destined to become a family heirloom. Another find along the quest for materials.
Photo 124 Air-source heat pump (air-to-air). External fan unit.
Photo 125 Air-source heat pump (air-to-air). Internal unit.
Photo 126 Another major overdraft contributor – three daughters who love musicals: farm conversion versus seats to see Hamilton!
Photo 127 The Tardis. In this case, the reclaimed shed did not actually fit into it.
Photo 128 The Tardis … and a few of the ultimate occupants..
Photo 129 Riding shotgun on the boiler: at least I had a heater.
Photo 130 Our offencing buzzard.
Photo 131 Our even more offending bottom.
Photo 132 Henry the Hoover.
Photo 133 Country living. When the road floods, it floods.
Photo 134 Country living. Bless the police for trying to divert cars, but this queue goes back 200m.
Photo 135 Country living. Definitely not a lane for a rear-wheel drive.
Photo 136 Country living. At the end of the day; many sunsets like this with no light pollution.
Photo 137 Country living. Kune Kune pigs; nightmares on legs.
Photo 138 Country living. Free-roaming chickens: fun, productive and friendly. Much safer than geese!
Photo 139 Country living. A large pond; a good and inexpensive groundworks investment. We did not add fish so have attracted many birds.
Photo 140 Country living. A year 20 overhaul. It took about six months for the wild grasses to re-establish.
Photo 141 Country living. We are delighted when the first pair of greylags appeared. For us the first day of spring.
Photo 142 Country living. We now have eight returning pairs (some shown). A bit noisy but they fall asleep before we do.
Photo 143 Country living. The roe deer in the woods are rather elusive: our eco-friendly reclaimed iron deer moved down from the north.
Photo 144 Postscript 1994. Memories of Sutherland where it all began in earnest. The first of the major projects began because of our love of salmon fishing and a dream to build a home close to the river. Farmhouse. Reusing the existing bathroom fittings posed no problem and looks fine.

List of plans

Plan 1 Mill. Initial outline plan of the Mill building (not to scale) with banked up earth completely obscuring one window and leaving only a lintel of the second exposed.

Plan 2 Mill. Final approved outline plan of the Mill building (not to scale) with part of the banked up earth removed showing the two lower floor windows and the new extension to the building where the original lean-to had been.

Plan 3 Mill. Drawing (not to scale) showing the existing layout of the lower floor of the Mill building.

Plan 4 Mill. Drawing (not to scale) showing the final approved layout of the lower floor of the Mill building.

Plan 5 Mill. Drawing (not to scale) showing the initial layout of the upper floor of the Mill building. This was submitted along with the planning proposal to show the lean-to.

Plan 6 Mill. Drawing (not to scale) showing the final approved layout of the upper floor of the Mill building.

Plan 7 Garage. Drawing (not to scale) showing the final approved lower floor of the building.

Plan 8 Garage. Drawing (not to scale) showing the final approved upper floor of the building.

Plan 9 Farm complex. Drawing (not to scale) showing an overview of the farm buildings with the arrangement of all of the buildings. The larger and smaller Barn areas are shown as Barn area 1 and Barn area 2.

Plan 10 Farm complex. Drawing (not to scale) showing the existing layout of the lower floor of the farm buildings. The larger and smaller Barn areas are shown as Barn area 1 and Barn area 2.

Plan 11 Farm complex. Drawing (not to scale) showing the final approved layout of the lower floor of the farm. buildings. The larger and smaller Barn areas are shown as Barn area 1 and Barn area 2.

Plan 12 Barn. Drawing (not to scale) showing the existing layout of the upper Barn areas (gin gan not shown as it had no upper area). There is a single mezzanine over about half of the area of the smaller part of the Barn became the Farmhouse kitchen. The larger and smaller Barn areas are shown as Barn area 1 and Barn area 2.

Plan 13 Barn. Drawing (not to scale) showing the final approved layout of the Barn with a second mezzanine created above the larger Barn area accessed by the new relocated staircase. The larger and smaller Barn areas are shown as Barn area 1 and Barn area 2.

Plan 14 Byre. Drawing (not to scale) showing the initial layout of the Byre.

Plan 15 Byre. Drawing (not to scale) showing the final approved layout of the Byre.

Plan 16 Farmhouse Extension. The position of the Farmhouse Extension (boxed in red) showing where it sits against the original farm layout, in almost exactly the same position as in the early aerial photograph (Photo 18).

Plan 17 Barn. Drawing (not to scale) showing the initial layout of the lower floor of the Barn Conversion.

Plan 18 Barn. Drawing (not to scale) showing the final approved layout of the lower floor of the Barn Conversion.

Plan 19 Barn. Drawing (not to scale) showing the existing layout of the upper floor (originally Barn area 1) of the Barn Conversion.

Plan 20 Barn. Drawing (not to scale) showing the final approved layout of the upper floor (originally Barn area 1) of the Barn Conversion.

Old Buildings: Conversion and Restoration

References

1. **Checkatrade** www.checkatrade.com
2. **Historic England** www.historicengland.org.uk/images-books/archive/collections/aerial-photos/
3. **National Collection of Aerial Photography** (NCAP) www.ncap.org.uk
4. **AnyVan** www.anyvan.com
5. **Shiply** www.shiply.com
6. **UK Pallet Commercial Deliveries** www.ukpalletcommercialdeliveries.com
7. **ParcelCompare** www.parcelcompare.com/courier-services/pallet-delivery
8. **Energy Saving Trust** www.energysavingtrust.org.uk
9. **Renewable Energy Website article: Roof Mounting of Wind Turbines** www.reuk.co.uk/wordpress/wind/roof-mounting-of-wind-turbines/
10. **Homebuilding & Renovating article: How to get Rid of Rubble** www.homebuilding.co.uk/advice/how-to-get-rid-of-rubble

Old Buildings: Conversion and Restoration

Further reading

Period Property Manual: Care and Repair of Old Houses
By Ian Alistair Rock
Publisher: Haynes Publishing Group, 2015
ISBN 978-0857338457

Renovating Old Houses: Bringing New Life to Vintage Homes
By George Nash
Publisher: Tauntcn Press, 2003
ISBN: 978-1561585359

The Victorian House Manual: Care and Repair for all Popular House Types
By Ian Alistair Rock
Publisher: Haynes Publishing Group, 2015 ISBN: 978-0857332844

The Vintage House: A Guide to Successful Renovations and Additions
By Mark Alan Hewitt and Gordon Bock
Publisher: WW Norton & Company, 2011
ISBN: 978-0393706192

The Old-House Journal: Guide to Restoration
Edited by Patricia Poore
Publisher: Dutton Adult, 1992
ISBN: 978-0525935513

Old Buildings: Conversion and Restoration

Please check out my website:
www.myoldbuildings.com

You can also find me on Facebook:
@myoldbuildings

I am happy to discuss the projects and invite you, my readers, and others to discuss renovation projects and personal experiences with period buildings on my Facebook page.